Results and Problems in Cell Differentiation

A Series of Topical Volumes in Developmental Biology

13

W0111511

Editors

W. Hennig, Nijmegen and J. Reinert, Berlin

Germ Line – Soma Differentiation

Edited by W. Hennig

With 34 Figures

Springer-Verlag
Berlin Heidelberg GmbH

Professor Dr. Wolfgang Hennig
University of Nijmegen, Department of Genetics
Faculty of Sciences
Toernooiveld, 6525 ED Nijmegen
The Netherlands

ISBN 978-3-662-21958-4   ISBN 978-3-540-39838-7 (eBook)
DOI 10.1007/978-3-540-39838-7

Library of Congress Cataloging in Publication Data.
Germ line – soma differentiation. (Results and problems in cell differentiation; 13). Includes bibliographies and index.
1. Cell differentiation. 2. Germ cells. 3. Somatic cells. I. Hennig, Wolfgang, 1941– . II. Series. [DNLM:
1. Cell Differentiation. 2. Germ Cells – cytology. 3. Hybrid Cells – cytology. W1 RE248X v. 13/ WQ 205 G3724]
QH607.R4   vol. 13   574.87′612 s [574.87′612]   86-10145 [QH607]

This work is subject to copyright. All rights are reserved, whether the whole or part of the material is concerned, specifically those of translation, reprinting, re-use of illustrations, broadcasting, reproduction by photocopying machine or similar means, and storage in data banks.
Under § 54 of the German Copyright Law, where copies are made for other than private use, a fee is payable to the publisher, the amount of the fee to be determined by agreement with the publisher.

© by Springer-Verlag Berlin Heidelberg 1986
Originally published by Springer-Verlag Berlin Heidelberg New York in 1986
Softcover reprint of the hardcover 1st edition 1986

The use of registered names, trademarks, etc. in this publication does not imply, even in the absence of a specific statement, that such names are exempt from the relevant protective laws and regulations and therefore free for general use.

2131/3130-543210

*The authors and the editor wish
to dedicate this volume to the memory of*

SIGRID BEERMANN

*in appreciation of her
pioneering cytological studies of
chromatin elimination in Copepods*

Preface

One of the oldest problems in developmental biology is the differentiation between germ line and somatic cells. The continuity of germ line cells between subsequent generations of multicellular organisms was first suggested by Owen, and later elaborated by A. Weismann to his famous germ line theory. His additional assumption that cellular differentiation was based on a differential representation of the genetic material in somatic cells was soon disproved. In some, apparently exceptional, cases, however, such differences in the genetic material between germ line and somatic cells were discovered. The best-known example is the nematod *Ascaris*. Boveri discovered and studied the fundamental differences in the karyotypes of germ line and soma of *Parascaris equorum*. Later, similar situations were found in some other organisms. However, in particular the work of Spemann demonstrated that cellular differentiation in general is not accompanied by fundamental changes of the genetic material. Subsequently, the relatively few examples of germ line-soma differences achieved by chromatin elimination processes have been considered as a curiosity. Experimental studies have been essentially restricted to *Ascaris* species and to the pioneering cytological studies of chromatin elimination by S. Beermann.

Despite the large proportions of the genome involved in chromatin elimination, our knowledge of this process is still very restricted. In particular the biological meaning of this differentiation process is entirely obscure. In this context one must, however, consider that also for the majority of DNA sequences in eukaryotic genomes the biological relevance is unclear. It is therefore not unlikely that an understanding of the biological role of chromatin elimination might be indicative for the biological role of those DNA sequences not belonging to the protein coding DNA fraction.

In this volume germ line-restricted parts of the genome are considered from different points of view. It is attempted to give a representative account of the known biological features of this portion of the genome. In a final chapter I have tried to provide some ideas which might correlate germ line-restricted DNA sequences to a more universal component of eukaryotic genomes, heterochromatin, and to introduce some ideas on the possible biological role of this part of the genome.

I should like to thank the authors of this volume for their commitment and collaboration.

Spring 1986   WOLFGANG HENNIG

Contents

The Differentiation of Germ and Somatic Cell Lines in Nematodes

By H. TOBLER (With 14 Figures)

Unusual Chromosome Movements in Sciarid Flies

By S.A. GERBI (With 12 Figures)

Molecular Reorganization During Nuclear Differentiation in Ciliates

By G. STEINBRÜCK (With 8 Figures)

Heterochromatin and Germ Line-Restricted DNA

By W. HENNIG

The Differentiation of Germ and Somatic Cell Lines in Nematodes

H. TOBLER [1]

1 Introduction

The segregation and differentiation of the germ and somatic cell lines in animal embryos represent one of the basic events in cell and developmental biology and has attracted classical and contemporary researchers alike. The process of germ line-soma segregation singles out cells which have to maintain the genealogy and those which build up the individual organism. This event often constitutes the first sign of differentiation in early cleavage stages of a developing embryo. Understanding, in molecular-genetic terms, such a basic developmental process might give us some clues to solve the most important problem in developmental biology, i.e., how cells become different during development. Although this review will focus on a discussion of germ line versus soma differentiation, it should be emphasized that the underlying mechanism(s) involved in the differentiation of the divergent somatic cell lines may in principle be similar.

Early cytogenetic research in nematodes carried out in the late part of the nineteenth century and the beginning of this century contributed enormously to elucidate fundamental cytological and biological phenomena (see Triantaphyllou 1971, for review). Thus, van Beneden (1883) discovered the process of meiosis in the maturing egg of *Parascaris equorum*. Furthermore, he and Boveri (1888) introduced in cell biology the concept of individuality and physical continuity of the chromosomes by showing that *Parascaris* chromosomes persist during the interphase stages of early cleavage divisions. Nussbaum (1880) and especially Weismann (1885, 1892) propagated the germ line theory, stating that there is continuity between germ cells of succeeding generations and that the germ line and the somatic cells segregate and develop independently of each other. As will be discussed in Section 2, it was the great cell biologist Theodor Boveri, who first demonstrated in 1887, by careful cytological analysis, that the egg of the nematode *Parascaris equorum* gives rise to a presumptive germ line and a presumptive somatic cell, thus confirming the germ line theory. Boveri also discovered the process of chromatin diminution in early cleavage stages of *Parascaris equorum* (see Chaps. 2 and 5). Nematodes provided further the first example of pseudogamy or pseudofertilization as a normal physiological strategy of reproduction in animals (Krüger 1913).

Given the many important classic contributions we owe to cell biological studies with nematodes, it is not surprising that these worms still play an important

[1] Institute of Zoology, University of Freiburg, Pérolles, 1700 Freiburg, Switzerland.

Results and Problems in Cell Differentiation 13
Germ Line – Soma Differentiation (Ed. by W. Hennig)
© Springer-Verlag Berlin Heidelberg 1986

role in present-day research. In *Caenorhabditis elegans*, for example, the entire developmental cell lineage from the zygote to the adult organism has been worked out. We know when the various somatic cell lineages separate from the germ line cells during early cleavage divisions; we further know the fate of each cell, including which somatic cell, at what developmental stages, is programmed to die (Sulston et al. 1983). Moreover, *Caenorhabditis elegans* is also well characterized from a genetic, cellular, and molecular biological point of view. Those nematodes which undergo chromatin diminution during early cleavage divisions, such as, e.g., *Ascaris lumbricoides*, are, of course, of special interest with respect to the problems of germ line versus soma differentiation (see Sects. 2, 3, and 5–10).

The purpose of the present article is to summarize the classic data on germ line-soma segregation and differentiation, to give a survey of some important concepts in developmental biology, and to disuss the earlier results in the light of recent and new findings which have been obtained using modern cell and molecular biological techniques. Special emphasis will be placed on the occurrence, possible mechanism, and function of the chromatin diminution process, as well as on the organization, structure, and function of the germ line-limited DNA sequences. Other phenomena which lead to genomic alterations in specific cells of eukaryotic organisms will also be briefly discussed in relation to chromatin diminution and chromosome elimination. Because the DNA elimination process occurring in ciliates, copepods, and sciarids will be reviewed in the following chapters of this volume, I will restrict myself to a discussion of nematodes, as far as the specific event of chromatin diminution is concerned.

2 The Germ Line Theory

2.1 Weismann's Germ Plasm Theory

According to Wilson (1896, cited in Eddy 1975), it was Owen (1849) to whom priority should be given for suggesting that the germ cell line (i.e., germ cell lineage or germ track) was continuous from one generation to the next, and that not all the progeny of the primary germ cell were involved in forming the body, but that some remained unchanged and contributed to the development of a new individual. Nussbaum (1880) was the first to state definitely that there was a continuity between germ cells of succeeding generations. He furthermore postulated that the cells concerned with the preservation of the species (i.e., the germ line cells) have a common origin with the cells forming the individual, but that the two cell lines divide independently of each other, giving rise to their proper progeny (see Hilscher 1983). However, it was August Weismann (1834–1914), the leading cytologist of the time, who made the theory of the germ line (in German: Keimbahn-Theorie) well known. He concluded that rather than continuity of germ cells there was continuity of a substance passed on from parent germ cells to the germ cells of the daughter individuals. He called this substance germ plasm, believed that it was located in the nucleoplasm, and postulated that it was responsible for the transfer of hereditary characters from generation to generation. He logically concluded that, if the germ and somatic cell lines remain independent of each other, then the hypothesis of the inheritance of acquired characters could not be upheld.

Weismann (1885, 1892) also offered a somewhat complicated hypothesis of ontogenetic development. He proposed that each chromosome which remains intact in successive generations of a species carries all the hereditary elements needed to make a whole individual. Because the different chromosomes of an individual may have been derived from many different ancestral lines, they must differ among themselves. Every single chromosome would potentially be able to build a whole organism, but in reality only one of the many chromosomes would be effective in determining a particular part of the body. Moreover, each chromosome was supposed to be composed of smaller subunits which Weismann termed biophores. The essential point of the intricate theory of development was that these biophores were distributed *unequally* to the somatic daughter cells during embryonic development, thus explaining cell differentiation by somatic gene segregation.

2.2 The Significance of Boveri's Discovery of Chromatin Diminution

The process of chromatin diminution in *Parascaris equorum* was first described by Boveri in 1887, and according to Baltzer (1962), this problem fascinated and intrigued him for over 20 years. As Fig. 1 shows, the first cleavage division of the *Parascaris* zygote cell proceeds normally, giving rise to two daughter cells, each containing two large chromosomes. During the second divisions of the cells, the chromosomes in the ventral P_1 cell are again normally distributed to the two daughter cells P_2 and S_2. This contrasts with the situation in the dorsal S_1 cell, where the central region of the chromosomes becomes fragmented. Only

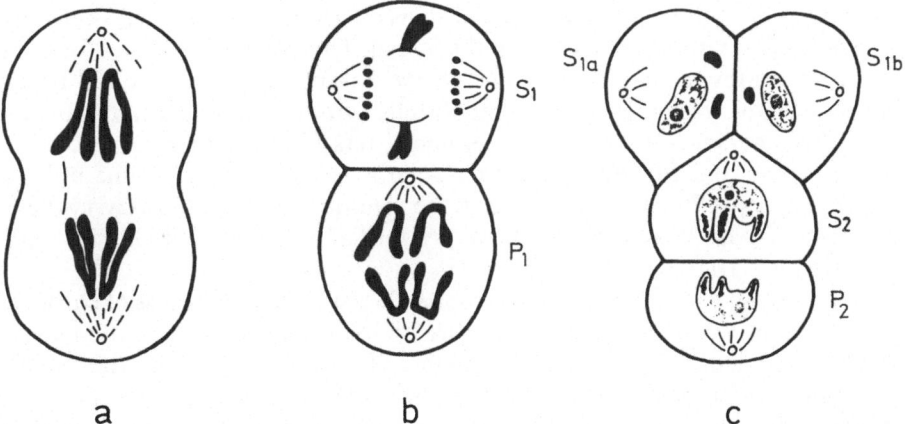

a   b   c

Fig. 1. Early cleavage divisions and chromatin diminution in *Parascaris equorum*. **a** Anaphase of 1st cleavage division. **b** Anaphase of 2nd cleavage division. Chromatin diminution takes place in the top S_1 cell but not in the lower P_1 cell. **c** 4-cell stage after completion of the 2nd cell division. The cells S_{1a}, S_{1b} and S_2 give rise to all somatic cells, whereas the P_2 cell represents the presumptive primordial germ cell. (Tobler 1976, after Boveri 1899)

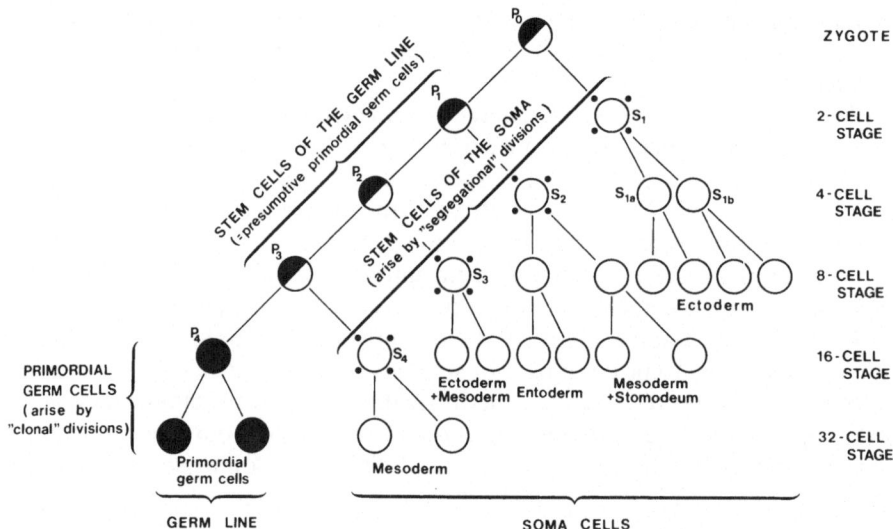

Fig. 2. Schematic representation of the cell lineage and segregation of germ line and somatic cells in *Parascaris equorum*. The presomatic cells S_1–S_4 undergoing chromatin diminution are indicated by $\bigcirc$. Final segregation of germ and somatic cell lines is achieved in the 16-cell stage in which the P_4 cell yields exclusively germ line cells. (Tobler et al. 1985, after Boveri 1910b)

these diminished chromosomes are distributed to the two daughter nuclei, the distal ends remain in the cytoplasm, where they eventually degenerate. Thus the two nuclei of the S_{1a} and S_{1b} cells contain less chromatin than the nuclei of the cells S_2 and P_2. The elimination of part of the chromosome and the concomitant fragmentation into a number of small chromosomes has been termed *chromatin diminution* by Herla (1893).

During the following round of cell division from the 4- to the 8-cell stage, chromatin diminution takes place in the S_2 cell. The process continues for a total of four divisions (cf. Fig. 2, cells S_1–S_4). By careful analysis of the cell lineage, Boveri (1910b) was able to show that all cells with the reduced amount of chromatin become somatic cells, whereas nuclei retaining the original integrity of chromosomes and full quantity of chromatin give rise to germ line cells (see Fig. 2). The P_4 cell therefore represents the primordial germ cell (PGC), whereas cells P_0 to P_4 are termed presumptive primordial germ cells (pPGC; for definitions see Sect. 3.1).

With regard to the germ line theory of A. Weismann, the process of chromatin diminution in *Parascaris equorum* confirmed the early segregation and independent development of the germ and somatic cell lineages. Moreover, the retention of the full genetic inventory in the germ line cells and the apparent elimination of a considerable amount of chromatin from the presumptive somatic cells was in line with Weismann's germ plasm theory. However, later experiments carried out by Boveri, his collaborators, and others disproved this latter aspect of Weismann's theory, namely, that gene segregation represents the basic mechanism of cell differentiation (see Sect. 5).

The chromatin diminution process in *Parascaris equorum* constitutes not only the first proof for the validity of the germ line theory in its broad sense, but allowed tracing the cell lineage of the germ line cells back to very early cleavage stages. This was accomplished by the fact that the germ line cells were traceable in the interphase of the cell cycle by their larger nuclei (due to their full chromatin content), compared to somatic nuclei, as well as during mitosis by the clearly differing chromosome morphology (cf. Fig. 1). One should bear in mind that the germ line theory does not predict that the segregation of germ line versus somatic cells takes place right during the first cleavage division of the zygote, or very early in development. The theory does, however, postulate that the segregation of germ line and somatic cells precedes all somatic determinations, as has been found to be the case in different animal phyla (see Sect. 3). Finally, chromatin diminution served to prove that a given region of the cytoplasm does exert a decisive influence on the behavior of the nucleus, as will be discussed in Sect 6.

2.3 Phylogenetic Aspects of the Establishment of the Germ Line

The segregation of germ line and somatic cells is phylogenetically very old and represents probably the first step in differentiation of a multicellular organism. This statement is based on the following line of evidence. The multi-cellular green alga *Volvox* sp. (see Fig. 3) is composed of only two different cell types: somatic cells which are morphologically álike and perform all the vegetative functions of the organism and the generative cells (i.e., oocytes and spermatozoa) which serve the organism's reproduction. The two cell types are morphologically and functionally different from each other; therefore a *Volvox* sphere is considered to be a multicellular individual and not just an aggregation of similar cells, e.g., a colony. Because *Volvox* represents a phylogenetically very old species which might even serve as a model of how multicellular organisms have arisen from unicellular organisms, the presence of morphologically and functionally, clearly

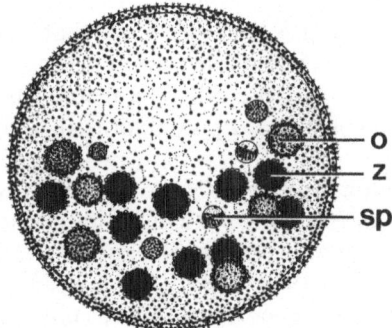

Fig. 3. *Volvox globator*, a many-celled green alga, which is composed of about 10,000 vegetative and about 20 generative cells. The vegetative cells are somatic, whereas the generative cells represent part of the germ line cell lineage and form either oocytes (o) or spermatozoa (sp). Zygotes (z) are indicated as *filled dark circles*. (Streble and Krauter 1982, by kind permission of Franckh'sche Verlagshandlung W. Keller and Comp., Stuttgart)

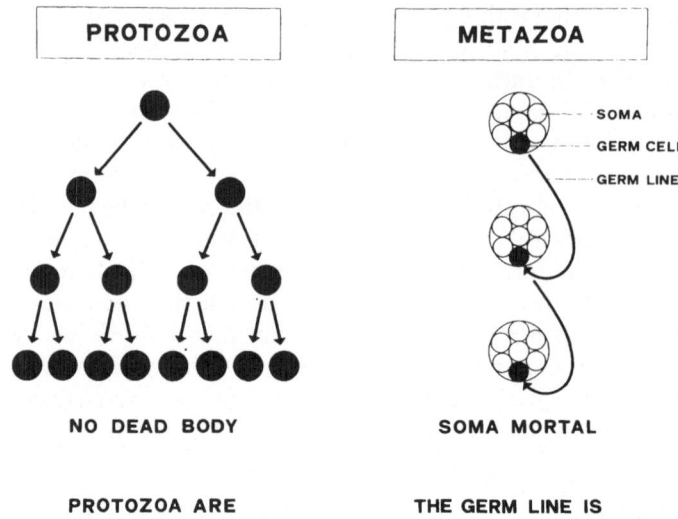

PROTOZOA

METAZOA

SOMA

GERM CELL

GERM LINE

NO DEAD BODY

SOMA MORTAL

PROTOZOA ARE
POTENTIALLY IMMORTAL

THE GERM LINE IS
POTENTIALLY IMMORTAL

Fig. 4. Comparison of the succession of generations in protozoa and metazoa. The schematic representation shows that protozoa and the germ line of metazoa are potentially immortal. Further explanations see text

distinct germ line and somatic cells supports the idea of an early segregation of the two cell lines during evolution.

The concept of the germ line adds a new dimension to the problem of immortality, if one compares the actual situation in protozoan and metazoan species. As presented in Fig. 4, there is no separation of germ and somatic cell lines in protozoa. Consequently, a single cell has to perform all the necessary functions of an individual. Following reproduction, there is no dead body left behind. In a metazoan species, however, the germ and somatic cell lines become segregated, leaving only the cells of the germ line potentially immortal, whereas the soma is destined to die. The comparison of the succession of generations in protozoa and metazoa, as depicted in Fig. 4, clearly shows that the fate of the germ line in metazoa corresponds to that of whole cells in protozoa. Looking at the problem this way, the death of the somatic cells in a metazoan species can be considered the consequence of their restricted potentialities acquired during cellular differentiation.

In the animal kingdom, early segregation of germ line and somatic cells during embryonic development is phylogenetically widespread. It occurs in several species of nematodes, arthropods, vertebrates, and others (for review see Wolff 1964, Nieuwkoop and Sutasurya 1979, 1981, McLaren and Wylie 1983). Even ciliates show a sort of germ line-soma segregation, in as much as the micronuclei are considered to represent the germ line equivalent, and the macronuclei the vegetative or somatic equivalent, respectively (see Steinbrück this Vol.). Because ciliates, nematodes, arthropods, and vertebrates do not represent closely related animal phyla, one has to assume that during evolution the segregation of germ line and somatic cells originated polyphyletically. As we will see in Section 5 of this chapter, the same argument may also be applied to the occurrence of chromatin diminution and chromosome elimination in the animal kingdom. It should be added that no germ line could be detected in some animal phyla (sponges, coelenterates, flatworms). This does, of course, not preclude the possibility that early segregation of germ line and somatic cells during ontogeny represents a common biological principle which has been abandoned in several animal phyla.

It is generally thought that no germ line exists in the plant kingdom. However, Müntzing (1949) in the diploid grass species *Poa alpina* and Darlington (1956) in *Sorghum purpurea-sericeum* described a behavior of accessory or B chromosomes which led Darlington to speak of germ line and soma. In both cases, B chromosomes are regularly perpetuated in the stem (which might be regarded as the germ line of a plant), and equally regularly lost in the root and other somatic tissues. Darlington (1956) considered these events to be analogous to the process of chromatin diminution known to occur in many animal species (see Sect. 5.2).

3 Segregation of Germ Line and Soma Cells

3.1 Definitions

In species where the germ line cán be traced through almost the entire life span of an individual, i.e., from fertilization to the actual germ cell formation, at least four different phases may be distinguished. Unfortunately the various authors have not always applied the same terminology. For the purpose of this review, we will use the terminology proposed by Nieuwkoop and Sutasurya (1981), but based on the definitions given by Rieger et al. (1968). The term germ cells, as opposed to somatic cells, comprises presumptive primordial germ cells (pPGC's), primoridal germ cells (PGC's), gametocytes, and gametes. The four cell types follow each other successively during ontogeny and represent the germ line (see Fig. 5). The pPGC's are the forerunners of PGC's, but they still give rise to both the germ line *and* somatic cells. With respect to the distribution of germ plasm to the daughter cells, they divide unequally. PGC's are defined as cells which yield *only* germ cells. They have a common clonal origin and allocate the germ plasm to both daughter cells. To give a specific example: the P_3 cell during early cleavage divisions of the *Parascaris* zygote is still a pPGC, whereas the P_4 cell represents a true PGC (see Fig. 2). We define the moment of segregation of germ line and soma cells as that specific developmental stage at which genuine PGC's are formed.

The pPGC's and the PGC's belong to the developmental period before sexual differentiation sets in (cf. Fig. 5). With the onset of sexual differentiation in the gonadal anlage of gonochoristic species, male and female individuals can be distinguished morphologically. The gametocytes (oogonia, spermatogonia) multiply, and some of them differentiate through meiosis to form functional gametes (ova, spermatozoa).

For the term germ plasm, I will adopt the definition given by Eddy (1975). He defines germ plasm as "a substance present in the cytoplasm of gametes (oocytes),

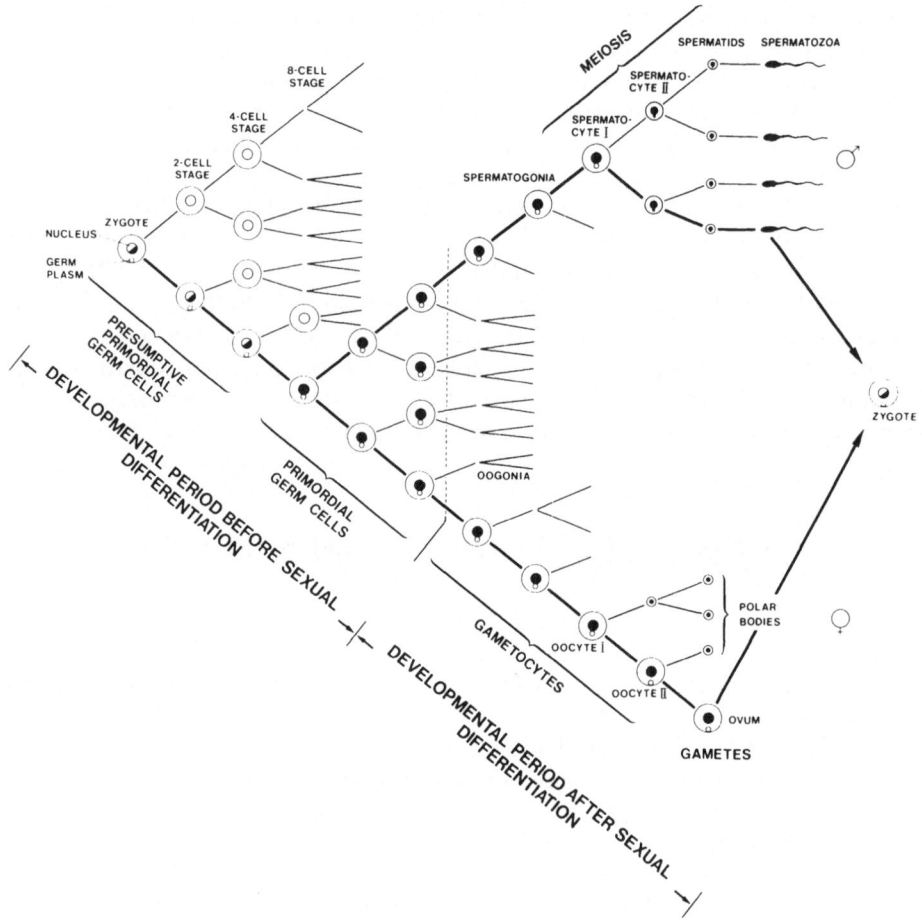

Fig. 5. Schematic representation of the successive phases in the development of the germ cells. The germ line is indicated by *thick lines*. Further explanations see text

which is segregated into specific cells (during blastulation) and determines that those cells shall become the 'progenitors of the germ cell line' (= PGC's) during subsequent development''.

3.2 Origin of the Germ Line and Experimental Confirmation of Germ Line Determination

Eggs of several animal species contain at their vegetative or posterior pole a particular cytoplasmic constituent called germ plasm (for review see: Beams and Kessel 1974, Eddy 1975, Mahowald et al. 1979). Synonymous terms that have been used are germinal plasm, germinal cytoplasm, germinal granules, germ line granules, dense bodies, polar granules, and polar plasm. Because these constituents of cells – at least in some favorable species of insects and amphibians – are present in all stages of the germ cell lineage, one has to conclude that the germ line originates from the vegetative or posterior pole of the egg. Moreover, germ plasm is also suited for establishing the liaison between succeeding generations of a species.

Ultrastructural analysis of the germ plasm of a variety of species revealed that form and distribution of these granules are quite similar in invertebrates and vertebrates. The germ plasm shows an irregular mass of dense fibrous or fibrogranular material with no surrounding membrane (see Mahowald et al. 1979, Eddy et al. 1981, for review). Similar structures have been found in PGC's of many animals, including mammals, and were termed nuages (see Eddy 1975).

Germ plasm was already thought to be specific for germ cell development by Hegner (1911), because destruction of the posterior cytoplasm in the egg by cauterization resulted in the formation of sterile adults. The same result was achieved by ultraviolet irradiation of the egg's polar region prior to nuclear immigration (Geigy 1931). By transplanting posterior polar plasm to the anterior pole or midventral region of a host embryo during early cleavage stages of *Drosophila* embryos, Illmensee and Mahowald (1974, 1976) demonstrated that the formation of cells with the morphological characteristics of pole cells was induced in an otherwise somatic region. This result has been confirmed by Ueda and Okada (1982). Upon further transplantation of such pole cells to the posterior pole of a second host, the cells differentiated in the gonad to yield functional gametes. Because genetic markers have been used to label donor, first, and second hosts differently, the results demonstrated that posterior polar plasm has the potential to induce the formation of PGC's in a presumptive somatic region.

In *Drosophila*, at least, cytoplasm of the posterior region of the egg must contain a germ cell determinant which functions autonomously upon transplantation to an ectopic region, whose prospective fate is to form somatic tissues. Is there any proof that the germ plasm which is located in the posterior polar region of the *Drosophila* egg is in fact the germ cell-determining constituent? Unfortunately not, because attempts to isolate and purify the germ plasm have so far failed. It can be said, however, that the presence of the germ plasm is associated with the germ line cells. This has been shown to be true for a number of species from different animal phyla (see reviews cited above).

3.3 Ontogenetic Aspects of Germ Line-Soma Segregation

Segregation of germ line and soma takes place in a cell if the two daughter cells turn out to be differently programmed, one giving rise exclusively to germ line and the other one exclusively to somatic cell lineage. Or, using the terminology introduced in Section 3.1, a presumptive primordial germ cell divides into one PGC and one stem cell of the soma. In the animal kingdom, the segregational process of germ line and soma occurs in various ways. As shown in Fig. 6, one single PGC (e.g., in *Parascaris equorum*) or one single nucleus (e.g., in *Mayetiola destructor*) may be ancestral to all future germ cells. Alternatively, the cell lineage of PGC's cannot be traced back to a single cell or nucleus, but rather to several cells or nuclei, as is the case in *Xenopus laevis* and *Drosophila melanogaster*. The future germ

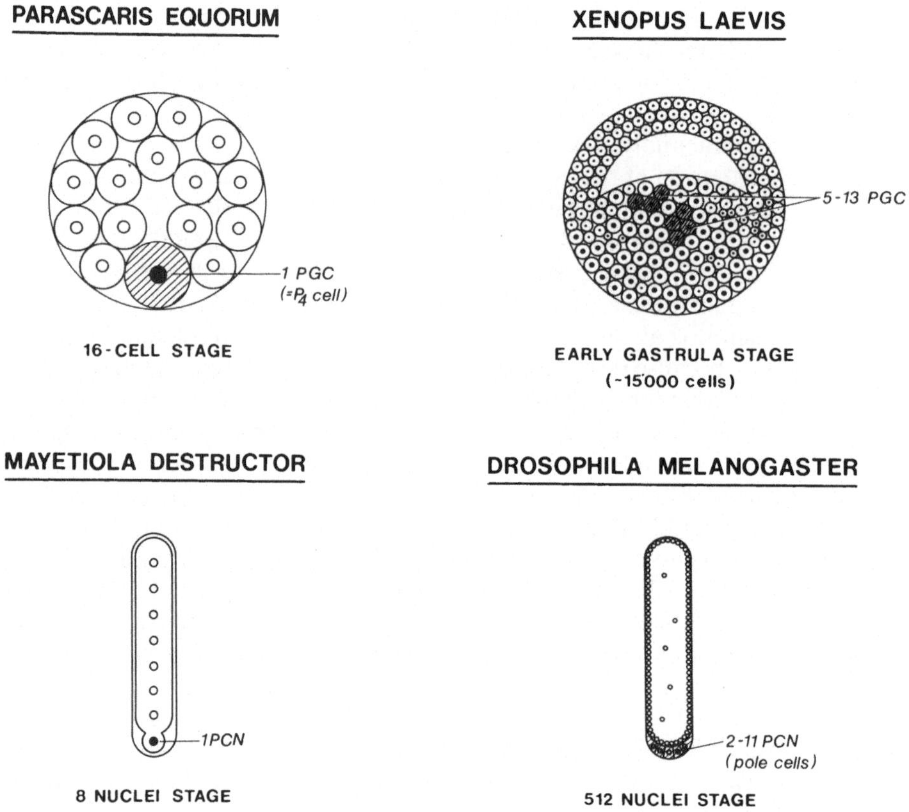

H. Tobler

PARASCARIS EQUORUM

16-CELL STAGE

1 PGC
(=P₄ cell)

XENOPUS LAEVIS

5-13 PGC

EARLY GASTRULA STAGE
(~15'000 cells)

MAYETIOLA DESTRUCTOR

1PCN

8 NUCLEI STAGE

DROSOPHILA MELANOGASTER

2-11 PCN
(pole cells)

512 NUCLEI STAGE

Fig. 6. Segregation of germ line and somatic cells in four different species. The cell lineage of PGC's can be traced back to one single cell in the 16-cell stage of *Parascaris equorum* (Boveri 1892) or to one single primordial cell nucleus (PCN) in the 8-nuclei stage of *Mayetiola destructor* (Bantock 1970). In *Xenopus laevis* and in *Drosophila melanogaster*, the cell lineage of the PGC's cannot be traced back to one single cell or nucleus, but to 5–13 PGC's in the early gastrula of *Xenopus* (Whitington and Dixon 1975) or to 2–11 pole cells in *Drosophila* (Sonnenblick 1965). The number of cells in the early gastrula of *Xenopus laevis* are taken from Gerhart (1980). (Tobler et al. 1985)

cells are therefore not always linear descendants from a single, specific, stem cell or nucleus.

Species-specific differences are also noted if one compares the various periods in the development of an organism in which germ line-soma segregation takes place (cf. Beams and Kessel 1974). Whereas germ line determination occurs early in some invertebrates (e.g., *Ascaris*, *Mayetiola*), this is not evident in many vertebrates until a much later period of development (e.g., *Xenopus* or the mouse). As a general rule, species showing an unequivocal germ line segregate early in ontogeny. This raises the question whether species with no apparent germ line segregate only very late so that an actual germ line goes undetected. The question may be academic, but I think this idea has to be rejected if one accepts the definition of the germ line given in Section 2. According to this definition, the germ line hypothesis postulates that (1) the germ line and somatic cells *segregate*, (2) develop *independently* of each other and that (3) germ line-soma differentiation *precedes* all somatic differentiation. This is clearly not the case in a wide range of invertebrates (e.g., coelenterates, flatworms), in which germ cells arise from undifferentiated pluripotent reserve cells, equipped with the capacity to differentiate into a number of somatic cell types in addition to germ cells. For example, the interstitial cells (I-cells) of coelenterates are the precursors of cnidoblasts, nerve cells, and germ cells (see, e.g., Tardent 1978). In this context it is interesting to note that I-cells of the coelenterate *Pelmatohydra robusta* contain electron-dense bodies, ultrastructurally and topographically similar to the germ plasm found in other animals (Noda and Kanai 1977). When the I-cells differentiate into germ cells, the total area of "germ plasm" increases about four times; however, when the I-cells differentiate into cnidoblasts, they decrease in number and size and eventually disappear. The direction of differentiation of I-cells into either oogonia (germ cells) or cnidoblasts (somatic cells) may be influenced by temperature. Therefore, environmental factors rather than germ plasm itself determines the direction of differentiation of the pluripotent I-cells (Noda and Kanai 1977).

Owing to the segregation and independent development of the germ line and somatic cell lineages in animals endowed with a germ line, these cells cannot be replaced by somatic cells, and sterilization by selectively killing the germ line cell lineage is therefore practicable (see, e.g., Eddy 1975). This is, of course, not the case in animals like coelenterates or planarians, which contain pluripotent stem cells throughout their body and therefore show a strong ability to regenerate. Moreover, they have the potential to produce offspring asexually, and the offspring possess germ cells.

3.4 Possible Mechanism for Germ Line Determination

As has already been mentioned, morphological observations as well as experimental findings are consistent with the hypothesis that germ plasm determines the germ cell line (see Dixon 1981, for review). This concept is supported by demonstrating that in the amphibian embryo for example, the amount of germ plasm is positively correlated with the resulting number of PGC's (Tanabe and Kotani 1974, Whitington and Dixon 1975, Wakahara 1978). The direct proof, however, that the germ plasm itself actually specifies the PGC's or germ line cells is still lacking. This would require the physical isolation and purification of germ plasm and its functional testing by injection into sterilized embryos. Attempts to isolate and analyze the germ plasm biochemically have been carried out in different laboratories (Wakahara 1977, 1978, Waring et al. 1978, Ueda and Okada 1982). From the vegetal pole cytoplasm of amphibian eggs, Wakahara (1977) isolated a heat-stable, nondialyzable, lyophilizable and UV-sensitive material which was able to restore the formation of PGC's when injected into UV-irradiated eggs. The PGC-forming activity was recovered from a 15,000 *g* pellet fraction isolated from the vegetal pole cytoplasm and was shown to contain some aggregates of electron-dense, minute granules similar to germ plasm, a large amount of membranous structures and some mitochondria, but no large yolk platelets. Because the ability to induce supernumerary PGC's was not abolished by digesting the fraction with pronase, Wakahara (1978) concluded that the PGC-determining activity should be associated with some material other than protein. This is, of course, in line with earlier UV-irradiation data by Smith (1966), who has shown that there is a positive correlation between the absorption spectra of nucleic acids and the UV-sterilizing effect. Earlier cytochemical studies by Blackler (1958) have already indicated that RNA is a major component of germ plasm because this plasm stained with Pyronine and because digestion with RNAse prevented staining. Similarly, Mahowald (1971 a) has shown that polar granules of *Drosophila* eggs contain RNA, which disappears during early stages of embryogenesis. He and his associates (Waring et al. 1978) also succeeded in isolating a "particulate subcellular fraction that by EM analysis consists predominantly of polar granules". Furthermore, they have identified a basic protein with a molecular weight of about 95,000 which seems to be unique in cell populations containing pole cells. However, there is so far no definitive proof that this protein really is a constituent of polar granules or germ plasm, nor is anything known about its function. Recently, Ueda and Okada (1982) isolated from a homogenate of young *Drosophila* embryos a subcellular fraction, which upon injection into UV-sterilized embryos was capable of inducing pole cell formation. The pole cell-inducing activity was mainly localized in the precipitate of a 27,000 g fraction. Dialysis, lyophilization and heating to 80 °C did not inactivate the pole cell-inducing capacity. Such pole cells, however, failed to develop into germ cells, thus indicating that this subcellular fraction is devoid of an important germ cell-determining factor which must be present in the polar plasm of normal eggs. Their observations indicate that germ cell determination requires at least two factors, one being UV-sensitive and one UV-resistant, but both seem to be localized at the posterior pole of the egg.

Despite the great efforts to isolate germ plasm physically from different animal species, success has so far been very limited. Because ultrastructural analysis of germ plasm and granules of a great variety of species indicated that they are not surrounded by a membrane, isolation of these cell organelles might be extremely difficult. Moreover, fractionation procedures applied so far may have caused germ plasm to dissociate into small granules and other components (see, e.g., Ueda and Okada 1982).

Following the localization of the germ plasm during ontogeny, Bounoure (1939) and Blackler (1958) found that in the amphibian egg it can be traced to small islands underneath the cellular membrane of the vegetative pole. The germ plasm could then be followed through early cleavage stages, blastula, gastrula, and neurula, and eventually to the gonadal rudiments of the larvae. Interestingly, at late blastula or early gastrula stages, the germ plasm becomes translocated from a peripheral location in the cell to a juxtanuclear position. This developmental stage coincides with the transition of pPGC to PGC's (see Sect. 3.1), or in other words, as long as the germ plasm is located marginally, it will be distributed unequally to the two daughter cells, hence the pPGC's still give rise both to germ line and somatic cells. However, once the germ plasm covers the cell nucleus, both daughter cells receive germ plasm and therefore both will give rise to germ cells.

In *Drosophila melanogaster*, the life cycle of the polar granules during ontogeny has been carefully analyzed by Mahowald (see Mahowald 1971 b, for review). The polar granules appear during vitellogenesis at the posterior tip of the oocyte and become attached to mitochondria. After fertilization, they are freed of mito-chondria, usually fragment, and acquire clusters of ribosomes at their periphery. As soon as pole cells have developed, the number of ribosomes decreases and reaggregation of the fragments occurs until the end of the blastoderm stage, when large aggregates have formed, no longer showing ribosomes. During the differ-entiation of the pole cells into PGC's, i.e., at the stage where the embryonic gonad is formed, the polar granules fragment again, become attached to the nuclear en-velope, and remain there until the next cycle of oogenesis. Although the develop-mental period at which germ plasm becomes closely associated with the nuclear membrane seems not to be exactly identical in amphibians and insects, it is inter-esting that the same alternation between a distant and a close arrangement of germ plasm and nucleus takes place during the ontogeny of both classes of ani-mals.

Germ line granules have also been observed in the fertilized egg and in the germ line cells of early cleavage stages of the nematode *Caenorhabditis elegans* (Krieg et al. 1978, Wolf et al. 1983). They segregate asymmetrically during the 1st and up to the 4th division of the germ line stem cells (pPGC's), but symmetrically during the 5th round of cell divisions in the P_4 cell (PGC), which gives rise exclu-sively to cells with pure germ line quality (Kimble and Hirsh 1979). Again, as in amphibians, the granules show a strong association with the nuclear membrane of the P_4 blastomere, although they have been dispersed into numerous small par-ticles in the cytoplasm of earlier stages (Wolf et al. 1983). By using fluorescent antibody staining, Strome and Wood (1982) reported the presence of cytoplasmic granules which are unique to germ line cells throughout the life cycle of *Caenor-habditis elegans*. Subsequently, monoclonal antibodies were used to identify the germ line specific cytoplasmic granules (Yamaguchi et al. 1983, Strome and Wood 1983). These elements were designated P granules and shown to become localized in the cytoplasmic region of the presumptive primordial germ cells prior to mitosis. Making use of mutants which cause abnormal spindle orientations during early cleavage divisions, and applying various microtubule and microfila-ment inhibitors, Strome and Wood (1983) further suggested that the asymmetric localization of the P granules is dependent on the proper function of microfila-ments. Because number, size, and distribution of the P granules correlate with those of germ line granules, it is very likely that the antiserum against the P factor recognizes some component of the germ line granules. However, and quite unfor-tunately, the antiserum does not react with the germ line granules after aldehyde fixation needed for electron microscopy, thus preventing so far a direct compari-son of the two types of granules (Wolf et al. 1983).

The most important question is, of course, what is the role of germ plasm? As we have already seen, proof for the hypothesis that germ plasm determines the germ cell line is still lacking. If it is established that germ plasm serves to segregate germ and somatic cell lines during development, then it will be interesting to know how this is performed in molecular-genetic terms. Two alternative ways of action have been proposed. The first one would be a negative influence or passive deter-mination step, with germ plasm acting as a protector of the nucleus from becom-ing determined to somatic cell lineages by cytoplasmic determinants (Blackler 1958; and recently restated by D. D. Brown 1984). This could be accomplished by somehow physically protecting the nucleus or by blocking a specific fragment of the genome. One of the weaknesses of this hypothesis is that UV-irradiation of eggs and early cleavage stages indicate that RNA rather than protein plays an important role in germ line determination, whereas proteins are generally thought to be better suited to act as protectors or repressors. The second hypothesis postulates a positive influence or active determination step, with germ plasm being engaged in germ line determination, perhaps by activating a specific segment of the genome. Mahowald (1968) has suggested that a maternal mRNA may be stored in the germ plasm during oogenesis. This mRNA would direct the synthesis of specific proteins which accumulate in sufficient quantity at the posterior pole of the egg to determine the entering cleavage nuclei for becoming pole cells. Some support for this hypothesis arises both from ultrastructural and cytochemical studies (see Mahowald 1971 b, for review). Moreover, Zalokar (1976) showed by autoradiographical analysis that in the blastoderm stage RNA is synthesized in the presumptive somatic cells, but not in the pole cells. During the same developmental stage, however, pole cells are heavily engaged in protein synthesis, much more so than presumptive somatic cells. This differential behavior of the two cell lineages could be the result of translation of stored mRNA in the pole cells at the blastoderm stage. The weakness of the hypothesis is that it is full of gaps and that so far no hard facts are available which prove that a mRNA specific for germ line determination is really stored in the posterior polar cytoplasm of the egg and early embryo.

Another possible function of the germ plasm has been attributed to the prevention of chromosome or chromatin loss in the germ line cells of those species which undergo chromatin diminution or chromosome elimination. I will discuss the evidence for this functional aspect of germ plasm in Section 6 of this article.

4 Mosaic Versus Regulative Development

4.1 Definitions

Around the turn of this century, it was realized by early experimental embryologists that animal eggs could be separated into two groups with respect to the developmental potencies of individual blastomeres. In the first type of egg, designated mosaic, an isolated blastomere forms only that part of an embryo which it would also form during normal development. In the second type of egg, called regulative, upon surgical isolation a blastomere is capable of giving rise to a whole embryo. Whereas mosaic development therefore is characterized by an early restriction of the developmental potentialities of individual blastomeres, this is not true for regulative development. Individual blastomeres may remain totipotent up to a given stage and become developmentally restricted only later during cleavage. By transplanting small tissue fragments of prospective epidermis into an ectopic region of the early amphibian gastrula, Mangold (1923) has shown that these cells regulate to conform to their new position. After gastrulation, however, the same transplanted tissue fragment no longer regulates, but forms structures which correspond to the site of their origin. The transplantation experiment therefore revealed that during the gastrula stage the tissue in question has become determined to form structures, which they would also have formed at their original location. According to Hadorn (1965), determination is defined as a "process which initiates a specific pathway of development by singling it out from among various possibilities for which a cellular system is competent". There has been much controversy whether the term determination really is a meaningful one in developmental biology. In my opinion, "determination" is a useful operational term, but it should only be applied if the following two requirements are met: it should always be stated (1) by what experimental criterion determination has been established and (2) to what structure(s) and/or function a given cell or tissue is determined.

Using the technique of surgically isolating early blastomeres and checking whether whole organisms are formed, one can state that determination takes place early in mosaic, but late in regulative development. How the determination process sets in during early cleavage divisions is not at all known, but it is generally believed that in mosaic embryos cytoplasmic determinative factors become segregated, whereas in regulative embryos differences between cells are established by cell interactions. Yet embryonic development is much more complex than that, and there is much evidence that both mechanisms and further ones are involved in the development of most species.

4.2 Experimental Analysis to Define the Type of Determination and Examples for Mosaic and Regulative Development

Fragmentation of early cleavage stages is not the only means to check whether a blastomere is still totipotent or not. Conversely, two or more early cleavage stages may be fused. If such combined embryos give rise to double monsters (as is the case, e.g., in *Sabellaria*; Hatt 1931), then the development of the egg follows the mosaic type. On the other hand, if such fused embryos are able to form normal-sized organisms (such as, e.g., amphibians, Spemann 1903, Mangold and Seidel 1927; or the mouse, Tarkowski 1961, Mintz 1962, Markert and Petters 1978), the development of the egg is said to follow the regulative type. A third method to check for the presence of mosaic or regulative development consists in setting a specific defect in early cleavage stages. This is accomplished by destroying or removing one or more blastomeres. Whether the egg will be classified as mosaic or regulative in its developmental character depends on the presence or absence of the respective lesions in the differentiated organism.

Applying one or more of the above-mentioned methods, it has been established that several species of ctenophore, nematode, ascidian, and mollusc follow a mosaic type of development. On the other hand, sea urchins, teleosts, amphibians, and mammals belong to the regulative type (see Wilson 1925, Kühn 1971, Davidson 1976, Slack 1983, and Sang 1984, for review). Some insects and annelids are classified as mosaic, others as regulative in their developmental character. Quite extreme cases are represented by the mammals on the one hand and by the nematodes on the other. In the mouse and rabbit, cells from the 8-cell stage can still contribute to both the inner cell mass (ICM), which forms the actual embryo, and to the extraembryonic trophoblast. Using chimeric combinations of marked cells, Hillman et al. (1972) showed that the differentiation of a blastomere into ICM or trophoblast depends on whether it is located on the inside or outside of the morula. There is so far no evidence of a cytoplasmic localization of a determining factor in the mammalian egg. This contrasts with the situation in the nematode *Caenorhabditis elegans*, where it has been demonstrated that the cell lineage during embryonic development is very rigid and that position and fate of each cell is determined by its ancestry. Not only is there a lineally transmitted, internally segregating, cytoplasmic component which is unique to the germ line cells (see Sect. 3.4), but there is also a somatic cytoplasmic marker ("rhabditin granule", Chitwood and Chitwood 1950) which could be traced from the gut cells of first-stage larvae all the way back to the 2-cell stage (Laufer et al. 1980, see Wood et al. 1983, for review). Cell lineage and cytoplasmic factors seem therefore to be of prime importance in the determination of a cell's fate in nematode embryology, rather than interactions with neighboring cells. There is some evidence that there are no unbound, prelocalized factors for determinate development in specific regions of the egg cytoplasm (Laufer and von Ehrenstein 1981).

It is interesting that species belonging to the mosaic or to the regulative type of development may contain germ plasm in their germ line cells (e.g., *Caenorhabditis elegans*, *Xenopus laevis*). But it is also true that species which so far have not been reported to contain germ plasm (e.g., ctenophores, sea urchins) may also belong to both developmental types. From the above it can be concluded that there is no correlation between the presence of germ plasm and the type of embryonic determination, i.e., mosaic or regulative development, in the animal kingdom.

5 Chromatin Diminution and Chromosome Elimination

5.1 Historical Review and Definitions

As already mentioned in Section 2.2, Boveri (1887) was the first to observe and describe the process of chromatin diminution in *Parascaris equorum*. Soon afterwards his student O. Meyer (1895) discovered that the same process occurred in the nematode *Ascaris lumbricoides*, *Ophidascaris filaria* and *Ascaris anguillae*. In 1908, Kahle reported that during early nuclear divisions of the young embryo of the fly *Miastor metraloas* (belonging to the dipteran family Cecidomyiidae), a considerable amount of chromosomal material is eliminated from the presumptive somatic nuclei. From his careful cytological observations he inferred that the chromosome number becomes reduced during the 3rd and 4th cleavage divisions, and that chromosomal ends are cast off at anaphase in the presumptive somatic nuclei. Kahle (1908) called this phenomenon "diminution process", undoubtedly with reference to chromatin diminution in *Parascaris* as described by Boveri. In 1934, Huettner (in *Miastor americana*) and Reitberger (in *Heteropeza pygmaea*) came to the correct conclusion that entire chromosomes rather than chromosomal fragments become eliminated from the presumptive somatic nuclei in the two Cecidomyid species analyzed. Therefore, the elimination process in the insects was called chromosome elimination as opposed to chromatin diminution in nematodes. Later cytological investigations revealed that *Miastor* possesses only 12 chromosomes in its somatic nuclei, in contrast to the germ cells, which normally contain 48 chromosomes (White 1954). Further research by many cytologists demonstrated that chromosome elimination occurs not only in several Cecidomyiid species, but also in five other systematic groups of insects which are phylogenetically quite unrelated (cf. Table 1).

Only relatively recently has chromatin diminution been detected in copepods and ciliates. S. Beermann (1959, 1977) demonstrated that such a process takes place in several *Cyclops* species (see Hennig this Vol.), and Ammermann (1965) was the first to show that loss of chromatin must occur during macronuclear formation in the ciliate *Stylonychia mytilus* (see Steinbrück also this Vol.).

With respect to some of the relevant definitions concerning the loss of chromatin and chromosomes, there is considerable confusion in the literature, which mostly results from an old definition of the term chromatin elimination which later on proved to be unfortunate. Seiler in 1914 introduced this term to denote the elimination of chromatin during the meiotic divisions of Lepidopteran oocytes. According to his cytological observations, relatively large amounts of material are shed from each chromosome at anaphase of the first meiotic division in the form of a definite body which remains in the equatorial plate as the daughter chromosomes move on to the poles. These "elimination bodies" degenerate later and ultimately disappear. The same process was cytologically confirmed to occur in several other Lepidoptera (see White 1973, for review), in Trichoptera (Klingstedt 1931), in Coleoptera (Rempel and Church 1969), in the grass mite *Pediculopsis graminum* (K. W. Cooper 1939) and in the tardigrade *Hypsibius dujardini* (Ammermann 1967). The problem with the term chromatin elimination started when Bauer (1932) reported that the eliminated material is Feulgen-negative, since chromatin, at least by contemporary definition, must contain DNA, associated proteins, and RNA. The lack of Feulgen-stainability of the "elimination bodies" has been confirmed by several workers using different species. Moreover, Ris and Kleinfeld (1952), applying various cytochemical methods, suggest that the cast-off material consists essentially of ribonucleoprotein. On the other hand, there is one report of failure to detect RNA in the eliminated material of *Bombyx mori* (Vereiskaya 1975). To my knowledge, no biochemical analysis on the composition of the isolated elimination chromatin has so far been carried out. Because most reports consider the elimination chromatin to be composed of ribonucleoprotein and containing no DNA, I suggest that the term chromatin elimination introduced by Seiler (1914) should be replaced by the more neutral term ribonucleoprotein shedding. Elimination can then be used in conjunction with chromosome and chromatin and means the loss of whole chromosomes or chromosome segments from any stage of the cell cycle. The process inevitably leads to differences in the chromosome complement between different cell lineages. The term diminution, which stems from the Latin expression diminuere and means to diminish and to fragment, could then be used for the process of breaking up a chromosome into several fragments, as has been originally described by Boveri (1887, 1910b) in *Parascaris equorum*. In other words, I suggest that the terms diminution and fragmentation be used synonymously to describe the process of breaking up a chromosome into several pieces as observed, e.g., in early cleavage divisions of the *Parascaris* egg.

5.2 Occurrence of Chromatin and Chromosome Elimination

Table 1 lists all the animal species in which chromatin or chromosome elimination has been reported to occur. Not included in this compilation are cases of chromosome elimination in plants (see Nagl 1976, for review), particularly the selective elimination of chromosomes in plant hybrids (see, e.g., Finch 1983, Thomas and Pickering 1983). Moreover, also not listed are the experimentally induced chromosomal losses which have been observed in various combinations of somatic cell hybrids (see, e.g., Weiss and Green 1967, Croce et al. 1973). Table 1 makes it clear that chromatin or chromosome elimination takes place in taxonomically scattered groups. The phenomenon was shown in four different animal phyla, in which there are many arthropods, but only few vertebrate species represented. Apart from some special cases of chromosome elimination to be discussed later, most elimination events take place in the presumptive somatic cells during early embryogenesis. In this respect even the nuclear differentiation processes occurring in ciliates compare well with those of metazoa: Ciliates contain two nuclei, a vegetative macronucleus and a generative micronucleus which is responsible for the sexual events (see Steinbrück this Vol.). Because macronuclei are always descendants of micronuclei, one can consider the successive stages of micronuclei as representing the germ line. Moreover, macronuclei are always formed early in development, which compares well with the early segregation of germ line and somatic cells in metazoans showing chromatin or chromosome elimination. Thus the throwing out of chromatin from the developing macronucleus in ciliates that undergo chromatin elimination may well have the same function as the elimination processes in multicellular metazoans.

From the phylogenetically widespread occurrence of chromatin and chromosome elimination in the animal kingdom, one might conclude that this event arose independently a number of times during evolution. This idea is further supported by the fact that the elimination process differs in various aspects among the different species as will be discussed below. It should also be mentioned that the polyphyletic origin of chromatin and chromosome elimination parallels the widespread but not universal occurrence of early segregation of germ line and somatic cell lineages (see Sect. 2.3). Likewise, not all ciliates or nematodes undergo chromatin elimination. The ciliate *Paramecium aurelia*, for instance, shows no indication of chromatin loss during macronuclear formation (Sonneborn 1974, McTavish and Sommerville 1980, see also Steinbrück this Vol. for review). In nematodes, no chromatin diminution has been observed by cytological methods in *Strongylus paradoxus*, *Rhabdonema nigrovenosa* (Bonnevie 1902) and in at least 11 more nematode species (Walton 1974). Also, no chromatin diminution has been found in the free-living nematode *Caenorhabditis elegans*, using biochemical and molecular biological techniques (Sulston and Brenner 1974, Emmons et al. 1979).

Chromatin and chromosome elimination usually takes place during early cleavage stages in the presomatic nuclei, as has already been briefly mentioned. In different nematode species, chromatin elimination begins with one exception in the 2nd or 3rd and ends at the 5th to 8th cleavage divisions (cf. Table 1). The only exception concerns *Physaloptera indiana*, in which the elimination process appears to last from the 1st to the 3rd cleavage division (Goswami 1973). Because an elimination process during the first cleavage division would not lead to a difference between germ line and somatic chromatin, I suggest that this report should be confirmed. In copepods, chromatin elimination may take place during the 4th, 5th, 6th, or during two successive cleavage divisions, depending on the species (cf. Table 1). Similar variations are also noted in different cecidomyiid species. The elimination cycle of Sciaridae is rather complicated, because chromosomes are eliminated during spermatogenesis and early cleavage divisions, whereby more chromosomes are discarded from the presumptive somatic than from the germ line cells (see Gerbi this Vol.). Chromosome elimination in the primordial germ cells and in somatic cell nuclei has also been observed in the Chironomidae (cf. Table 1).

Some cases have been reported in which chromosomal material is not eliminated from presomatic cells, but rather during spermatogenesis (in *Macroceroea grandis* and *Hydrolagus colliei*), or during mitotic parthenogenesis (*Strongyloides papillosus*, cf. Table 1). In *Ctenocephalides orientis*, one of the two Y chromosomes becomes eliminated from male somatic cells, whereas in *Metaseiulus occidentalis*, half of the chromosomes are discarded from the germ line and probably also from most somatic cells in male embryos (cf. Table 1). These chromosome elimination processes may in some way be related to sex determination. In this respect, the chromosomal behavior in the Coccidae or scale insects is extremely interesting. The more primitive coccids show heterochromatization of the paternal set of chromosomes in early embryogenesis in the males, but not in the females (see S. W. Brown and Chandra 1977, or Nur 1980, for review). Since these condensed chromosomes are genetically inert, the males can be considered to be physiologically haploid, although they remain diploid with respect to their chromosome number. In the more advanced coccid species, the paternal chromosome set becomes eliminated and not just inactivated during early ontogeny of the male embryos, thus males develop as true haploids. There is evidence that the change from heterochromatization to elimination occurred several times (S. W. Brown and Chandra 1977, Nur 1980). The coccid system therefore tells us that chromosome inactivation by heterochromatization preceded chromosome elimination and that this succession evolved independently more than once during evolution, at least in these insect species.

Chromosome elimination has also been observed to occur in some marsupials (cf. Table 1). Females lose one of the two X chromosomes, males the Y chromosome in somatic tissues (Hayman and Martin 1965, 1969). Subsequent studies revealed that the pattern of chromosomal loss differs among tissues and species (see, e.g., Close 1984, for a most recent review). Because the casting out of one sex chromosome from somatic tissues in marsupials parallels the heterochromatization of an X chromosome in the females of other mammals, it has been suggested that the elimination process serves as a dosage-compensation mechanism

Table 1. Occurrence of chromatin diminution or chromosome elimination

Group		Species	Event	References
Protozoa:	Ciliata	– Stylonychia mytilus	During macronuclear formation	Ammermann (1965)
		– Euplotes aediculatus	During macronuclear formation	Ammermann (1971)
		– Oxytricha sp.	During macronuclear formation	Lauth et al. (1976)
		– Paramaecium bursaria	During macronuclear formation	Schwartz and Meister (1975), Steinbrück et al. (1981)
		– Trachelonema sulcata	During macronuclear formation	Kovaleva and Raikov (1978)
		– Tetrahymena thermophila[a]	During macronuclear formation	M.C. Yao and Gall (1979)
		– Tetrahymena pyriformis[a]	During macronuclear formation and during macronuclear division upon induced inhibition of DNA replication	Andersen (1972), M.C. Yao and Gorovsky (1974)
		– Chilodonella cucullulus	During 1st post-conjugation fission	Radzikowski (1973)
Nematoda:		– Parascaris equorum (= A. megalocephala)	During 2nd or 3rd–5th cleavage divisions	Boveri (1887, 1899)
		– Ascaris lumbricoides var. suum (= A. lumbricoides)	During 3rd–5th cleavage divisions	O. Meyer (1895), Bonnevie (1902)
		– Ophidascaris filaria (= A. rubicunda)	During 2nd or 3rd–5th cleavage divisions	O. Meyer (1895)
		– A. anguillae (= A. labiata)	?	O. Meyer (1895)
		– Contracaecum incurvum (= A. incurva)	During 3er–5th (?) cleavage divisions	Goodrich (1916)
		– Toxocara canis (= A. canis)	During 2nd–6th cleavage divisions	Walton (1917, 1924)
		– T. cati (= Belascaris mystax)	During 2nd or 3rd–6th cleavage divisions	Walton (1924)
		– T. vulpis (= Belascaris triquetra)	During 3rd–6th cleavage divisions	Walton (1924)
		– Cosmocerca sp.	During 3rd–8th (?) cleavage divisions	T. Yao and Pai (1942)
		– Physaloptera indiana	During 1st–3rd cleavage divisions	Goswami (1973)
		– Strongyloides papillosus[b]	During mitotic parthenogenesis giving rise to free living ♂	Albertson et al. (1979)
Crustacea:	Copepoda	– Cyclops strenuus	During 4th cleavage division	S. Beermann (1959, 1977)
		– C. furcifer	During 6th and 7th cleavage divisions	S. Beermann (1959, 1977)
		– C. divulsus	During 5th cleavage division	S. Beermann (1977)
		– C. varicans	?	S. Beermann (1959)
		– C. bohater	During 5th cleavage division	Einsle (1964)
		– C. vicinus	During 6th cleavage division	Einsle (1964)
		– C. abyssorum	During 5th cleavage division	Einsle (1964)
		– Canthocamptus staphylinus	?	S. Beermann (1959)
		– Acanthocyclops vernalis	During 4th and 5th cleavage divisions	Akifjew (1974)
Insecta:	Coccina	– Pseudaulacaspis pentagona	Elimination of the paternal chromosomes in ♂ during blastula stage	S.W. Brown and Bennett (1957), Bennett and S.W. Brown (1958) see White (1973)
	Cecidomyidae	– and others		
		– Miastor metraloas	During 3rd and/or 4th cleavage divisions	Kahle (1908)
		– M. americana	During 3rd and/or 4th cleavage divisions	Hegner (1914), Nicklas (1959)
		– Mayetiola destructor	During 5th cleavage division	Bantock (1961, 1970)

Group	Species	Event	References
	– Wachtliella persicariae	During 4th cleavage division	Geyer-Duszyńska (1959, 1966)
	– Monarthropalpus buxi	During 5th cleavage division	White (1950)
	– Heteropeza pygmaea (=Oligarces paradoxus)	During 3rd cleavage and 4th and 6th–8th divisions	Reitberger (1934, 1940)
	– Mycophila speyeri	During 4th cleavage division in pedogenetic ♀ eggs, 5th and 6th in pedogenetic ♂ eggs	Nicklas (1960), Camenzind (1971)
	– Taxomyia taxi	?	White (1947)
	– and others (at least 11 more species)		see Matuszewski (1982)
Sciaridae	– Sciara coprophila	During spermatogenesis and during 5th or 6th and 7th or 8th cleavage divisions in presomatic (more) and germ line (less) nuclei	DuBois (1932, 1933), Metz (1938)
	– S. impatiens		Crouse et al. (1971)
	– Rhynchosciara sp.		Basile (1966)
	– and others (about 20 more species)		see Gerbi (this Vol.)
Chironomidae (Orthocladiinae)	– Trichocladius vitripennis	During and after 6th cleavage division in the presumptive somatic cells and during 1st mitotic division in the gonads of freshly hatched larvae in the primordial germ cells	Bauer and W. Beermann (1952a, b)
	– Metriocnemus cavicola		Bauer and W. Beermann (1952b)
	– M. hygropetricus		Bauer and W. Beermann (1952b)
	– M. inopinatus		Bauer and W. Beermann (1952b)
	– Psectrocladius obvius		Bauer and W. Beermann (1952b)
	– and others (at least 11 more species)		see Bauer and W. Beermann (1952b)
Siphonaptera	– Ctenocephalides orientis[b]	Elimination of one of the two Y chromosomes from ♂ somatic cells	C. Thomas and Prasad (1980)
Heteroptera	– Macroceroea grandis[b]	Elimination of fragments of X during anaphase I of meiosis in ♂♂	Banerjee (1959)
Acarina:	– Metaseiulus occidentalis[b]	Half of the chromosomes eliminated from the germ line and probably also from most somatic cells in ♂ embryos (2n=6)	Nelson-Rees et al. (1980)
Vertebrata: Chondrichthyes	– Hydrolagus colliei[b]	Elimination of chromatin during spermatogenesis	Stanley et al. (1984)
Marsupialia	– Isoodon macrourus[c]	Elimination of the Y in ♂ and of one X chromosome in ♀ somatic tissues	Hayman and Martin (1965, 1969)
	– I. obesulus[c]		
	– Perameles nasuta[c]		
	– P. gunni[c]		
	– and others (at least 6 more species)		see Close (1984)

[a] Since only ≈ 10% of the sequences present in micronuclei seem to be absent in macronuclei, one cannot decide with certainty whether the difference is due to sequence elimination or underreplication.

[b] These species do not show the typical elimination of chromosomal material in presomatic cells and represent therefore special cases of chromosome elimination.

[c] Chromosome elimination has probably replaced chromosome inactivation, which represents the usual dosage compensation mechanism found in other mammals.

(Hayman and Martin 1965, 1969). Similar to the situation in coccids, chromosome elimination may be regarded as an extreme form of chromosome inactivation, although the process might serve an entirely different purpose. Further striking similarities between the undoubtedly related phenomena of chromosome elimination and chromosome inactivation by heterochromatization will be discussed in Section 10.5.

To my knowledge, almost no chromosome and chromatin elimination processes are known in plants, apart from the preferential elimination of B chromosomes from at least some somatic tissues in *Poa* and *Sorghum* (see Sect. 2.3), the elimination of univalent chromosomes during the second meiotic division of the pollen mother cell of *Rosa canina* (see Lima-de-Faria 1983) and apart from the throwing out of chromosomes in species hybrids and somatic cell hybrids. I would also like to draw the readers' attention to a recent brief report by Nagl (1983), in which he presents microscopic and cytophotometric evidence for the occurrence of chromatin elimination in the orchid *Dendrobium*. The casting out of chromatin takes place through the nuclear membrane of interphase nuclei in protocorms and leaves. The significance of this process, however, still remains obscure.

In reviewing all the reported chromatin and chromosome elimination events listed in Table 1, it becomes immediately clear that the process mostly takes place during differentiation of germ line versus somatic cells in presomatic cells, but may also occur in other instances at different developmental stages and in different tissues. The process can furthermore take place at various stages of the cell cycle, during mitosis, meiosis, or at the interphase stage. Whole chromosomes, such as, e.g., in *Mayetiola destructor* (Bantock 1961, 1970), or only portions of it (e.g., in *Parascaris equorum*, Boveri 1887, 1910b) may be discarded. Moreover, the entire haploid set can be thrown out (e.g., *Pseudaulacaspis pentagona*, S. W. Brown and Bennett 1957), just single autosomes (as, e.g., in *Sciara coprophila*, Metz 1938), or individual sex chromosomes (*Ctenocephalides orientis*, C. Thomas and Prasad 1980). The eliminated material may amount to more than 80% as in *Parascaris equorum* (Moritz 1970b, Moritz and Roth 1976; see also Sect. 7), or to as little as about 10% (e.g., in *Tetrahymena thermophila*, M. C. Yao and Gall 1979). Accordingly, the number of chromosomes involved in chromosome elimination varies also enormously among the different genera and species. It ranges from about 56 chromosomes in certain *Chironomus* species, where the germ line cells carry about 60 chromosomes compared to only two pairs of autosomes in the somatic cells (Bauer and Beermann 1952b), to one to three cast out chromosomes in some *Sciara* species (see Metz 1938). Even in relatively closely related species, as in *Sciara coprophila* and *S. ocellaris*, the difference may be important, in as far as the former species eliminates chromosomes from the somatic cells, whereas the latter species loses none (see Metz and Lawrence 1938, as well as Gerbi this Vol.). Finally, the nature of the chromatin which becomes eliminated may also vary to a great extent, as will be discussed in Section 8 and by Steinbrück (this Vol.).

5.3 Cytological Observations of the Chromatin Elimination Process in Nematodes

Because the cytology of chromosome fragmentation and elimination in ciliates, copepods, and sciarids will be reviewed in the following three chapters of this book, I intend to discuss in this section only the cytological observations on the chromatin elimination processes in the two nematode species *Parascaris equorum* and *Ascaris lumbricoides* (var. *suum*). The other nematode species studied thus far show either no demonstrable chromosome elimination (e.g., *Caenorhabditis elegans*, see Sect. 5.2), or they show chromosome elimination, but virtually nothing has been reported on the cytological events occurring during this process (e.g., in *Toxocara canis*, see Walton 1917, 1924).

As mentioned earlier, the phenomenon of chromatin diminution was first discovered in *Parascaris equorum* by Boveri (1887). His careful and detailed description and documentation of the process (cf. Fig. 7, Boveri 1899, 1910b) has since been reproduced in many textbooks and reviews. During the 2nd cleavage division of the dorsal S_1 cell, the central portions of the chromosomes fragment into many smaller ones, while the large chromosomal ends are eliminated and remain in the cytoplasm where they eventually degenerate (see Figs. 1 and 7). This phenomenon has been termed chromatin diminution by Herla (1893). In the other blastomere, the ventral P_1 cell, the chromosomes retain their original number, shape, and size. During the 3rd and 4th division of the germ line stem cell lineage,

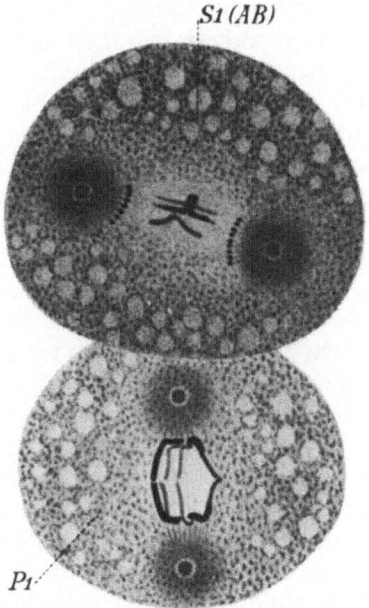

Fig. 7. Cleaving 2-cell stage of *Parascaris equorum* showing the fragmentation of the chromosome in the upper S_1 cell and the retention of the intact chromosome complement in the lower P_1 cell during anaphase stage. (Halftone copy of the original color illustration of Boveri's paper, published in 1899)

the same process is repeated; that is, the cell divides into one blastomere which undergoes chromatin diminution and one blastomere in which the chromosomes are undiminished (see Fig. 2). At the 5th cleavage division (P_4 cell), the germ line is completely segregated from the presumptive somatic cells, and chromatin diminution ceases. Boveri (1910 b) subsequently reported that chromatin diminution may begin only at the 3rd instead of the 2nd cleavage division, but then occurs simultaneously in cells S_{1a}, S_{1b}, and S_2. This was later confirmed by Fogg (1930) and Moritz (1967 a), who reported that diminution takes place during the third cleavage division in the race or variety *univalens* (n = 1), but may occasionally occur during the second cleavage division in the race *bivalens* (n = 2) of *Parascaris equorum*. The pattern of the diminutive divisions may also vary somewhat if one compares different nematode species (cf. Table 1).

In *Parascaris equorum*, the chromatin elimination process actually consists of two different events, namely of chromosome fragmentation and of the loss of chromosomal material from the presomatic cells. Fragmentation involves the thinner central portion of each chromosome and results in the formation of about 40 (Fogg 1930) to 70 (Kautzsch 1913) small individual chromosomes (see von Ubisch 1943). The large, unfragmented chromosomes carried in the germ line cells are therefore called collective or compound chromosomes (translated from the German expression Sammelchromosomen; see, e.g., Goldschmidt and Lin 1947). Elimination consists in casting off the thicker, club-shaped, end portions of the collective chromosomes into the cytoplasm during each diminuational division of the presomatic cells. These chromosome fragments degenerate later on in the cytoplasm and are probably resorbed. Painter and Stone (1935) and Schrader (1935) advanced the hypothesis that the collective chromosomes carried in the germ line cells of *Parascaris equorum* are peculiar in having a large number of spindle attachment sites or kinetochores instead of only one. To test this hypothesis, White (1936) irradiated fertilized eggs of *Parascaris equorum* with X-rays, which resulted in a partial fragmentation of the collective chromosomes. Such fragments, when equipped with a functional spindle attachment site, are distributed to the daughter nuclei during the first cleavage division of the P_0 cell. It seemed as if the spindle attachment sites are confined to about the middle third of the length of the collective chromosomes, thus leaving the end portions of them without their own spindle. Such a polycentromeric chromosomal organization would, of course, explain why the spindle-fiberless terminal parts of the chromosomes remain in the metaphase plate after chromosome fragmentation in the presomatic cells, and are not distributed to the two daughter nuclei. In a recent paper, Goday et al. (1985) presented ultrastructural evidence that the mitotic chromosomes in the germ line cells of *Parascaris equorum* var. *univalens* and *bivalens* are equipped with a very long continuous kinetochore structure, running along a considerable part of the whole chromosome. They further concluded that in *Parascaris equorum* var. *bivalens*, the kinetochore spans not only the euchromatic chromosomal regions, but also the intercalary heterochromatic portion of the chromosome. However, Goday et al. (1985) were unable to ascertain whether the kinetochore plate covers the whole chromosome, including the germ line-limited, terminal heterochromatic regions, although they suggest that this is the case. What seems to be clear from the studies of Goday et al. (1985), is that the collective chromosomes carried in the germ line cells of *Parascaris equorum* contain a continuous kinetochore (holocentromeric organization), rather than multiple, discrete, kinetochore units (polycentromeric organization). A holocentric chromosomal organization has already been inferred by Moritz (1967a), at least for the variety *Parascaris equorum univalens*. Using classical cytological techniques, he showed that spindle fibers are also present in the germ line-limited chromatin, although they are less abundant than in the intercalary chromosomal region, which is retained in the somatic cells. Furthermore, he observed in many preparations an anaphase activity of the terminal chromosomal parts in the germ line cells. These observations led Moritz (1967a) to question the existence of akinetic chromosomal ends in the collective chromosomes of the germ line cells. It seems to me, therefore, that the question of whether or not the germ line-limited chromosomal parts carry centromeres, and if so whether they are functional in germ line cells but nonfunctional in presomatic cells, is still open.

Bauer (1932) was the first to report that the eliminated chromatin is strongly Feulgen-positive. It must therefore contain a large amount of DNA. This was later confirmed by Goldschmidt and Lin (1947), Pasteels (1948a, b) and Lin (1954). Moreover, all these authors distinguished between the euchromatic and heterochromatic portions of the collective chromosomes. Whereas the former part is carried both in the germ line and the somatic cells, the latter seemed to be germ line-limited and therefore absent from all somatic cells. Microspectrophotometric determinations of Feulgen-stained metaphase chromosomes revealed that the DNA is also replicated in those cells which are destined to undergo a chromatin elimination cycle during the next cell division (Moritz 1967a). These quantitative analyses furthermore indicated that the amount of DNA in the collective chromosomes of the germ line is not constant, but varies to a large extent between different individuals, as well as between different spermatids of the same individual, at least in the variety *univalens* of *Parascaris equorum* (Moritz 1970a, b). Recently, Goday and Pimpinelli (1984) presented cytological evidence using Hoechst 33258 staining and a C-banding procedure that the heterochromatic chromosome regions are arranged differently in the two classical varieties *univalens* and *bivalens* of *Parascaris equorum*. Whereas the *univalens* variety contains only terminal heterochromatin in the collective chromosome, *bivalens* carries, in addition, some intercalary heterochromatin. Chromatin diminution in the presomatic cells leads to the elimination of all cytologically detectable heterochromatin in both varieties, regardless of its localization within the collective chromosomes (Goday and Pimpinelli 1984). Whether the two varieties *univalens* and *bivalens* of *Parascaris equorum* should be considered as two different species, i.e., *Parascaris univalens* and *P. equorum*, as has been proposed by Biocca et al. (1978) and Bullini et al. (1978), remains to be seen. Hybrids between the two varieties are regularly observed in large populations of early cleavage stages (see, e.g., Goday and Pimpinelli 1984). However, whether or not such hybrids are fertile and propagate in their hosts is not known. The suggestion that the two varieties of *Parascaris* represent two species was based on morphological and karyological characters, as well as on a series of electrophoretic studies of several enzyme loci, which revealed differences between the two *Parascaris* forms (Biocca et al. 1978, Bullini et al. 1978). Of course, the different organization of heterochromatin in the germ line

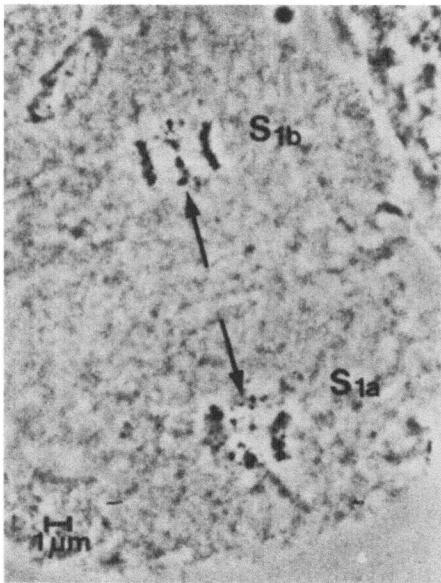

Fig. 8. Chromatin diminution in *Ascaris lumbricoides* var. *suum*. The elimination process takes place during the 3rd cleavage division in the two presomatic cells S_{1a} and S_{1b}. *Arrows* point to the cast off chromatin which during the anaphase stage remains in the equatorial plate. Part of the eliminated chromatin in the lower S_{1a} cell seems to be in the process of desintegration and resorption. (Felder 1983, unpublished)

chromosomes between the two varieties of *Parascaris* detected by Goday and Pimpinelli (1984) tends to support this view. In any case, it would be interesting to compare the localization of heterochromatin in the germ line chromosomes of *Parascaris equorum* var. *univalens* and *bivalens* with that of var. *trivalens*, which has been reported some time ago (Ju-Chi-Li 1937, Tchou Su and Chen-Chao-Hsi 1937) and which carries even six chromosomes in the germ line cells.

Chromatin elimination in *Ascaris lumbricoides* var. *suum* (see Fig. 8), the pig's intestinal parasite, is similar to that of *Parascaris equorum*. The elimination process, however, does not occur until the third and lasts up to the fifth cleavage division (O. Meyer 1895, Bonnevie 1902). Although it is beyond any doubt that the cell lineage patterns of germ line and somatic cells are invariant during early cleavage divisions of *Ascaris lumbricoides* var. *suum*, as is the case in *Parascaris equorum* (see Fig. 2) and *Caenorhabditis elegans*, there is evidence that the temporal sequence of the cleavage pattern may vary from the 4-cell to the 8-cell stage in *A. lumbricoides* (Goldstein 1977). Using whole mount staining of early cleavage stages, Goldstein (1977) was able to show that, for example, an S_{1a} cell in a 4-celled embryo may divide before, after, or concomitantly with the P_2 cell of the same generation. Secondly, no collective or compound chromosomes are present in the germ line cells, as is the case in *Parascaris equorum*, but rather single chromosomal units (Bonnevie 1902). Thus, during chromatin elimination, there is also no need to break up a few collective chromosomes into a large number of smaller chromosomes. Because the collective chromosomes of *Parascaris equorum* have often been considered to be composed of many tiny, spherical chromosomes, joined end to end (see, e.g., Lin 1954), it is likely that the *Parascaris* collective chromosomes arose by lateral fusion of small individual chromosomes during evolution. Alternatively, nematode species which have single chromosomes might have arisen from species with ancient collective chromosomes. In this context it is interesting to note that five different varieties of *Parascaris equorum* have been reported which contain haploid chromosome numbers ranging from 1–9 in their germ line cells (Walton 1959).

Attempts to determine the correct number of germ line chromosomes in males and females of *Ascaris lumbricoides* var. *suum* have produced conflicting results. Edwards (1910 a, b) reported male heterogamety with spermatids containing 19 and 24 chromosomes and haploid eggs with 24 chromosomes for *Ascaris lumbricoides*, so that the female karyotype would consist of $2n = 2(19A + 5X) = 48$ and that of males of $2n = 2 \times 19A + 5X = 43$ chromosomes. Different authors have found other chromosome numbers (cited in Vassilev and Mutafova 1974, and Goldstein and Moens 1976). Goldstein and Moens (1976), using 3-D reconstruction from electron microscopy of serial sections of male and female pachytene nuclei, determined 12 synaptonemal complexes in both sexes and concluded that n equals 12. They further suggested that the five heterochromatic masses present in the spermatocyte nuclei may represent five sex chromosomes. In a succeeding paper, Goldstein (1978) proposed a sex determination system for *Ascaris lumbricoides* var. *suum* in which the males are heterogametic, producing sperm with $n = 12$ and $n = 12 + Y_1 Y_2 Y_3 Y_4 Y_5$, respectively. The unfertilized eggs would all have a haploid chromosomal set of $n = 12$, thus giving rise to zygotes with either 24 or 29 chromosomes (Goldstein 1978). Vassilev and Mutafova (1974) and Mutafova (1975), however, reported that during meiosis, females form 24 bivalents and males 19 bivalents and 5 univalents, thus confirming Edward's early data. Moritz (1977) reported the same numbers of 48 chromosomes in female and 43 in male *Ascaris lumbricoides*. We were also able to confirm these values in our own laboratory (Felder 1983; see Fig. 9). I therefore have no doubt that the karyotype of *Ascaris lumbricoides*, originally reported by Edwards (1910 a, b), must be the correct one. Furthermore, I cannot believe that there exists a real difference in karyotype between the American and European subspecies of *Ascaris lumbricoides* var. *suum* because the domestic swine, and probably with it the endoparasitic *Ascaris* worm, has been brought to North America only relatively recently, so that there was hardly enough time for different karyotypes to evolve. Moreover, if one looks at Fig. 1 of Goldstein's paper published in 1978 and counts the chromosomes in the microphotograph representing a prometaphase stage of the first spermatocyte meiotic division, one sees in the polar view a central aggregate mass surrounded by 19 chromosomes. This is exactly what would be expected if during the meiotic prophase 19 autosomal bivalents are formed, whereas the 5 X-chromosomes aggregate to form a chromosomal mass.

Bonnevie (1902) reported that there is no fragmentation of the central portion of each chromosome during the chromatin diminution process in *Ascaris lumbricoides*. Because only the ends of the chromosomes would break off and remain in the cytoplasm, the number of chromosomes in both the germ line and somatic nuclei must be the same. Unfortunately, the germ line chromosomes in *Ascaris*

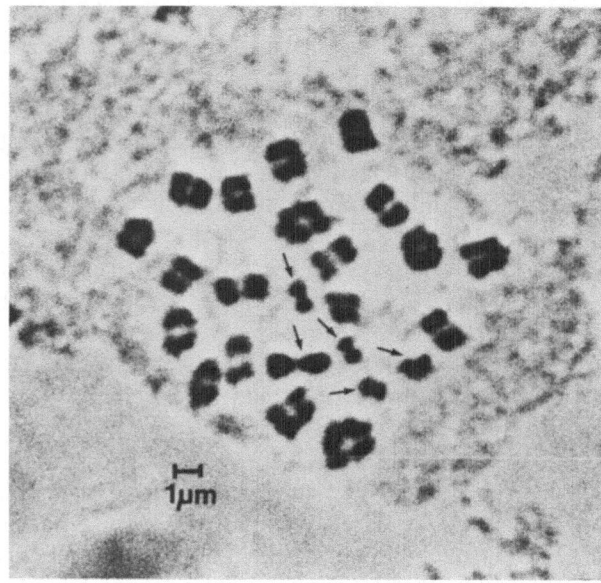

Fig. 9. Phase-contrast microphotograph of unstained spermatocyte I chromosomes of *Ascaris lumbricoides* var. *suum*. The spermatocytes are at the metaphase I stage of meiosis. Nineteen bivalent autosomes and 5 monovalent X chromosomes (*arrows*) are clearly seen. (Felder 1983, unpublished)

lumbricoides are very tiny (the largest one measures about 2 µm, on the average they are only about 1 µm long, cf. Fig. 9), thus after chromatin elimination they are even smaller, making chromosome counts in somatic cells extremely difficult. We are therefore not sure whether only terminally located chromatin is cast off during chromatin elimination, as Moritz (1984) claimed, or whether also intercalary chromatin is excised, which might increase the number of somatic chromosomes.

5.4 Cytoplasmic Factors and Cytochemistry

According to Weismann's (1885, 1892) and zur Strassen's (1896, 1898, 1906) theories, chromatin diminution results from factors which reside in the chromosomes themselves, thus the process would be autonomous and not induced by any cytoplasmic constituents. Boveri (1899, 1904, 1910b), on the other hand, postulated that differently distributed cytoplasmic factors localized in the *Ascaris* egg determine whether or not chromatin diminution occurs. As we will see in the next section, sophisticated experiments and careful observations allowed allocation of this determinative role to the cytoplasm. The question therefore arose whether there are visible inhomogeneities within the egg's cytoplasm and/or whether unequally distributed cytoplasmic factors can be detected using cytochemical techniques. Boveri noticed that the *Ascaris* egg is not homogeneous with respect to the distribution of the vitellin granules they are more concentrated at the vegetative pole (see, e.g., Nigon 1965). Because UV irradiation of the germ line cytoplasm at 260 nm wavelength allows chromatin diminution also to occur in the germ line chromosomes (see Sect. 6), it has been suggested that nucleic acids or their precursors are important in preventing chromatin diminution from taking place. However, Pasteels (1948 a, b), using cytochemical techniques, first demonstrated that granules containing ribonucleic acids are denser and coarser in the germ line cytoplasm than in the somatic cell line cytoplasm of the *Ascaris* egg. Moritz (1967 a) showed that a significantly elevated content of RNA in the germ plasm cannot be demonstrated before the second cleavage division. He therefore suggested that the excess of cytoplasmic RNA in the P_1 cell is not the cause, but rather the consequence of the germ line identity of this cell. Thus, still unknown factors must control the initial determinative steps at the beginning of development. We do not even know whether these factors are laid down already during oogenesis, or only after fertilization, or during both developmental periods.

With respect to the localization and distribution of other cell constituents such as glycogen particles, lipid droplets, refringent granules, mitochondria, and ribosomes in the oocyte, zygote, and early cleavage stages of *Parascaris equorum* and *Ascaris lumbricoides* var. *suum*, I refer the reader to the review by Anya (1976).

6 Functional Aspects of Chromatin Diminution and the DNA Sequences Restricted to the Germ Line

6.1 Experimental Analysis of Chromatin Diminution in *Parascaris*

We have just seen in the last section that Boveri (1899, 1904, 1910 b) advanced the hypothesis that localized cytoplasmic factors determine whether or not chromosomes undergo diminution. This idea was supported by a cytological analysis of dispermic eggs. Boveri (1910 b) found that occasionally two spermatozoa may enter and fertilize an *Ascaris* egg. If this happens, the egg usually cleaves simultaneously into four blastomeres. Such 4-cell stages may be divided into the following three types: Type I is composed of one presumptive primordial germ cell and three primordial somatic cells, type II forms two pPGC's and two primordial somatic cells, and type III forms three pPGC's and one primordial somatic cell (Fig. 10). The germ line or somatic character of each cell was established during the next cleavage division by verifying whether or not the chromosomes were diminished. If chromatin diminution did occur in a given cell, then all chromosomes were affected irrespective of their number. Boveri (1904) therefore concluded that the fate of a chromosome to become diminished cannot be an inherent property of the chromosome itself, but must depend on environmental (= cytoplasmic) factors. Because the first two cleavage planes in such eggs are randomly oriented, the blastomeres may or may not receive a given portion of the germ cell-determining cytoplasmic region of the egg, thus determining its fate (cf. Fig. 10).

In a further study of chromatin diminution in *Parascaris equorum*, Boveri (1910a) and Hogue (1910) found that in a centrifuged egg, diminution occurred

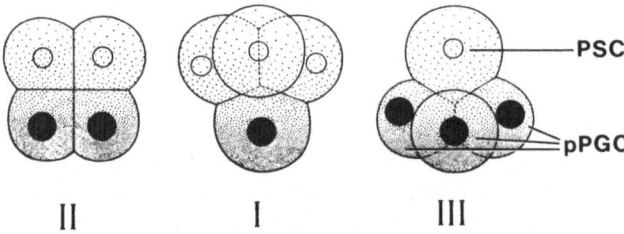

Fig. 10. Chromatin diminution in dispermic eggs of *Parascaris equorum*. Type I–III eggs denote the three different cleavage patterns as observed in dispermic eggs. In each case the cytoplasmic factors thought to determine germ line quality of the cells and distributed in a gradient-like fashion in the egg cells are represented by *fine dots*. *PSC* primordial somatic cell; *pPGC* preprimordial germ cell. (After Boveri 1899, 1910b)

only in specific regions, namely in proximity to the so-called ball which was produced in some of the eggs at the animal pole. Because the cleavage furrow often formed perpendicular to the normal cleavage plane in untreated eggs, the two blastomeres must have received roughly equal amounts of animal and vegetal cytoplasm. In the next division the two blastomeres adjacent to the ball undergo diminution, whereas the other two do not. The conclusion therefore was again that only those cells which have received the cytoplasmic region predestined to form germ cells retain the full complement of chromatin. In a recent communication, Moritz and Bauer (1984) reported that the development of the egg of *Parascaris equorum* var. *univalens* may reversibly be blocked by applying hydrostatic pressure. Thus, 2-cell stages may be formed in which one blastomere contains no, the other four, chromosomes. If all chromosomes are rendered to the presomatic cell S_1, diminution will take place in its two daughter cells. However, if all the chromosomes are distributed to the germ line precursor cell P_1, no diminution occurs in the P_2 daughter cell, but all four chromosomes undergo synchronous diminution in the daughter cell S_2. The results support the conclusion that the differential behavior of the chromosomes, i.e., nondiminution in the germ line, diminution in the somatic cells, depends on cytoplasmic factors and is not an intrinsic property of the chromosomes themselves. Because blastomeres containing no chromosomes, as well as their daughter cells, apparently cleave normally, postzygotic gene and mitotic chromosome activity seems not to be necessary for the maintenance of the orderly cleavage pattern during early embryogeny (Moritz and Bauer 1984).

Tadano (1968) reported the successful reversible conversion of germ line blastomeres into somatic blastomeres and vice versa by subjecting *Parascaris equorum* eggs to different chemical treatments. In eggs treated with lithium ions (e.g., LiCl), a substance known to have a vegetalizing effect in sea urchins and to cause a mesodermalization of chordomesoderm in amphibian embryos (see, e.g., Kühn 1971), the germ line blastomeres are induced to undergo chromatin diminution and resemble normal somatic blastomeres. On the other hand, chromatin diminution is inhibited in presomatic blastomeres treated with thiocyanate ions (e.g., NaSCN, an animalizing substance in sea urchin embryos, resulting in chorda hypertrophy in amphibians, see, e.g., Kühn 1971). Again, such cells resemble normal germ line cells. To my knowledge, up to now these interesting experiments have never been repeated, so that the rather surprising findings remain thus far unconfirmed.

UV-irradiation of *Parascaris* eggs resulted in the formation of various abnormal embryos (Stevens 1909, Schleip 1923, and others). In an extensive experimental study of *Parascaris equorum* eggs, Moritz (1967b) came to the conclusion that UV irradiation damage was largely due to inactivation of cytoplasmic RNA. Total UV irradiation caused chromatin diminution in all the blastomeres. The cells were more sensitive to irradiation at telophase than at any other stage of the cell cycle. Moreover, Moritz (1967b) presented evidence for the occurrence of an UV sensitivity gradient in the egg, exhibiting a peak in the region where the presumptive germ cells are formed. In summary then, all the observations and experimental results presented in this section support the conclusion that in the *Parascaris* egg, there exists a localized cytoplasmic substance, close to or at the vegetal pole, which becomes incorporated into the primordial germ cells, and whose function is to prevent chromatin diminution from occurring in all the germ line cells.

6.2 Hypothesis Concerning the Induction or Prevention of Chromatin Diminution in *Parascaris*

The notion that the uncleaved *Parascaris* egg contains a gradient of cytoplasmic factors, being more concentrated at the vegetal than at the animal pole, and whose function is to inhibit chromatin diminution, dates back to Boveri (1910b). The role of this cytoplasmic substance is in this case assumed to be a passive one. On the other hand, King and Beams (1938) proposed a hypothesis in which a substance would actively cause chromatin diminution to occur. Their hypothesis was based on findings from the following centrifugation experiments (King and Beams 1937, 1938). Uncleaved eggs of *Parascaris equorum* var. *univalens* were subjected to prolonged high speed centrifugation, whereby in most of the eggs cytoplasmic cleavage was suppressed, while nuclear divisions continued. They noticed that in such eggs diminution usually took place in all of the nuclei at the second or third nuclear division. This experimental result led them to postulate a diminisher substance D that is progressively produced from some cytoplasmic constituent An (which is spatially fixed and more concentrated at the animal than at the vegetal pole), and that actively causes chromatin diminution. The factor An changes slowly to D during development. Normally, D reaches sufficient concentration to induce diminution in the cells S_1 and S_2 at the second and third cleavage divisions, as well as in the presomatic cells derived from the presumptive primordial germ cells at succeeding cell divisions. In the absence of cell boundaries in the centrifuged egg, D is assumed to diffuse throughout the whole cytoplasm of the egg, and, upon reaching a sufficient concentration, to cause chromatin diminution in all nuclei.

Von Ubisch (1943) has criticized the hypothesis of King and Beams (1938) and pointed out that the region of the uncleaved egg, which later becomes incorporated into the first primordial germ cell P_4 (see Fig. 2), does not comprise the most vegetative material, but is rather located about midway between the equator and the vegetal pole. This piece of evidence has been brought to light by the three-dimensional reconstruction of the cleavage pattern and has apparently been overlooked both by Boveri and zur Strassen. Like Boveri, von Ubisch (1943) advanced the hypothesis of a passive role of a cytoplasmic substance on chromatin diminution, assuming that this cytoplasmatic constituent is localized in the lower part of the egg and inhibits diminution of chromatin. According to von Ubisch (1943), this substance reaches its highest concentration in a region of the egg which is located at 13/16th of the length of the animal-vegetal axis.

Pasteels (1948 b) proposed a novel hypothesis based on his cytochemical demonstration that ribonucleoproteins are unevenly distributed in the *Ascaris* embryo. Because chromatin diminution does not occur in cells which are strongly basophilic, possibly due to elevated contents of RNA, he suggested that ribonucleic structures might prevent chromatin diminution from occurring. Moreover, Pasteels (1948 b) assumed the existence of interactions between the two poles of the egg and the cortical structures. During the cleavage divisons, polar plasm becomes displaced, and he thought this to be an indication of rearrangement of cortical substances which might control the positioning of the spindle.

A further hypothesis has been advanced by Nigon et al. (1960). These authors have studied the behavior of the cell surface of several nematode species during early cleavage divisions. They proposed that the embryonic development is regulated by an interaction between the cell cortex, the positioning of the spindles, and certain cytoplasmic constituents. Cortical structures would yield the first elements of polarity. They furthermore suggested that the determinative factors for chromatin diminution are localized in the cortex rather than in the cytoplasm.

At the present time it is difficult to favor one or the other hypothesis concerning the induction or prevention of the chromatin diminution process in *Parascaris equorum*, given the almost complete lack of reliable biochemical data on the cytoplasmic constituents which are thought to be involved in chromatin diminution. What is really needed is the successful isolation and molecular characterization of cytoplasmic constituents which, using a functional assay, will prove to induce or prevent the chromosome fragmentation and elimination process. In the light of the present lack of such information, I personally favor the simplest hypothesis which is in line with all the experimental facts. It seems to me that it is still the classical hypothesis by Boveri (1910 b), as modified by von Ubisch (1943), which meets all these requirements.

6.3 Functional Aspects of Chromatin and Chromosome Elimination

With respect to the loss of genetic material from the presumptive somatic cells during the chromosome elimination process, the important question arises whether or not the germ line-limited DNA sequences have any function at all in the germ line cells and if so, what their exact function may be. Transplantation of somatic nuclei containing the reduced amount of DNA into enucleated eggs could give us an answer. Due to technical difficulties, however, such experiments have so far not been successfully performed in the *Ascaris* egg. Germ line cells are of course totipotent by definition; this totipotency is not necessary for somatic cells. In fact, in multicellular organisms there is no a priori reason why they should carry the full genomic information in their somatic cell lines. Every species showing chromatin or chromosome loss in the presomatic cells clearly proves that it contains in its germ line a fraction of chromatin which is not needed for the proper function of the somatic cells.

The best evidence for a functional role of the germ line-limited chromosomes still arises from some experiments carried out on two *Cecidomyid* species. Normal embryos of *Wachtliella* (Geyer-Duszyńska 1959, 1966) and *Mayetiola* (Bantock 1961, 1970) retain the full complement of about 40 chromosomes in the germ line nuclei, but lose about 32 chromosomes in the presumptive somatic cells. UV irradiation of the posterior pole of the egg prior to nuclear colonization led to the disintegration of the invading primordial germ line nuclei, so that no pole cells were formed. Somatic nuclei with the reduced number of 8 chromosomes later on replaced the germ line nuclei, thus forming new pole cells. Such embryos develop into sterile though otherwise normal adults. From this result one has to conclude that polar granules and the eliminated chromosomes are not directly necessary for pole cell formation, but are needed for normal gamete formation. In a further analysis of experimentally induced sterility, the egg was constricted at the 2- to 4-nuclei stage to prevent migration of cleavage nuclei into the posterior region. The constriction was removed after chromosome elimination in all nuclei had taken place, allowing one nucleus with the reduced number of chromosomes to invade the posterior pole of the cell, which still contained morphologically normal polar granules. Again, such experimentally treated embryos developed into normal but sterile adults. This result strongly suggests that posterior cytoplasmic components of the egg (most likely the polar granules), are responsible for preventing an irreversible loss of chromosomes from the invading presumptive germ line nuclei, but that they are not directly required for gamete formation. However, the chromosomes normally retained in the germ line cells seem to be indispensable for normal gametogenesis, particularly for oogenesis (Geyer-Duszyńska 1966, Bantock 1970). These experiments therefore tell us that the eliminated chromosomes contain not only "junk" DNA, but may have important functions for gamete differentiation and perhaps also for germ line determination.

6.4 Possible Function of the Germ Line-Specific Genetic Material

Although we have just seen that in *Wachtliella* and *Mayetiola* there is strong evidence that the germ line-limited chromosomes exert important functions during differentiation of the germ cells, I personally do not believe that this is necessarily true for *all* cases in which chromatin or chromosome elimination takes place in the animal kingdom (cf. Table 1). This opinion is based on the following considerations. First, chromosome elimination does not always segregate germ line and somatic cells (see Table 1). Second, the process is highly variable in its occurrence and mechanism (see Sect. 5.2). Such a variability is rather unlikely if the process were always important in germ cell differentiation. Third, chromosomal loss is a relatively rare event in the animal kingdom, and the species of those taxonomic groups in which it does occur usually constitute minorities. It is therefore rather unlikely that chromosome elimination serves as a general mechanism for the control of germ line differentiation. Fourth, the eliminated material is often heterochromatic in nature (see Sect. 5.3 for *Parascaris* and Hennig, this Vol., for *Cyclops*), which is suggestive of containing satellite or simple sequence DNA. Such DNA sequences are normally not transcribed (see Sect. 8), and it is difficult to see how they could play a major role in germ cell differentiation. We have therefore to consider further possible functions of the germ line-limited DNA sequences.

Boveri (1910b) already discussed three possible functions of the eliminated chromatin resulting from elimination in the presomatic cells of *Parascaris equorum*. He favored the idea that the germ line-limited chromatin was in some way of genetic importance for the development of the germ cells, because these sequences were obviously not needed in the somatic cells. In 1914, Seiler, while describing the chromatin elimination process in Lepidoptera, suggested the possibility that diminution may restore the balance between the nucleus and the cytoplasm (in German: Kernplasmarelation, after Hertwig 1903) in the maturing egg. Von Ubisch (1943) proposed that a specific cytoplasmic constituent is carried in the germ line cells protecting the germ line from becoming differentiated in a somatic direction. Moreover, this special cytoplasm later on ensures the correct differentiation of the primordial germ cells into the proper germ cells. A similar hypothesis has been advanced by Goldschmidt and Lin (1947). They considered the cast-off chromatin to be heterochromatic and genetically inactive, but proposed that the heterochromatin in the germ line cells is needed for continuous mitotic cell divisions, because nematodes are animals with a constant number of somatic cells. According to Painter (1945), the eliminated material would be functional in supplying nucleotides for the rapid cleavages of somatic cells that follow (and precede!) elimination. He later modified his hypothesis, stating that in *Parascaris equorum* the presence of germ line-limited chromatin serves to increase the polysome capacity of the oocyte during egg formation, whereas in the Cecidomyidae this is accomplished by the extra chromosomes of the nurse cells (Painter 1966). Kaulenas and Fairbairn (1968) suggested that the primary function of the germ line-limited DNA might consist in producing large amounts of RDNA in oogenesis or following fertilization, which during early cleavage become eliminated from the presumptive somatic cells. Wallace et al. (1971) modified this hypothesis somewhat, proposing that the eliminated material might have been carried along in the germ line cells as independent rDNA episomes. As we will see in Section 8.3, such a hypothesis is no longer tenable.

Further possible functions of the germ line-limited DNA sequences that have recently been advanced include the following. The cast-off chromatin (1) could be necessary for the control of gene expression by repressing the activity of those genes in the germ line cells which ought to be active only in somatic cells, (2) could have important functions in meiotic processes such as chromosome pairing and recombinational events, (3) might increase the volume of germ line nuclei, thus slowing down the duration of the cell cycle, (4) might represent a genetic reservoir carried in the germ line for evolutionary purposes, and (5) could have no function at all and represent useless, "junk" or selfish DNA. Of course, several of these and some of the aforementioned possible functions need not be mutually exclusive. I also believe that not all chromosome and chromatin elimination processes listed in Table 1 serve the same purpose. Thus far, we have no detailed knowledge about the meaning and the function of the elimination process or the germ line-limited DNA sequences in any of the species presented in Table 1. However, I think that from an elucidation of the genetic informational content of the eliminated DNA sequences as well as from an understanding of the mechanism of the elimination process in molecular-genetic terms, we will eventually be able to deduce the meaning and the function of the chromosome and chromatin elimination process. In the following Sections 7 to 9, we will discuss the molecular aspects of chromatin elimination in *Ascaris*.

7 Amount of Eliminated DNA

7.1 Evaluation of the Different Methods for the Quantitative Determination of Eliminated DNA

In principle, four different methods have been applied to determine the amount of eliminated DNA during chromosome and chromatin elimination in different taxonomic groups. These methods include Feulgen-microspectrophotometry, diphenylamine assay, isotope dilution and DNA renaturation kinetics. In order to account for the quantity of eliminated DNA, the amount of nuclear DNA before and after elimination has been determined, so that the difference should represent the amount of cast-off DNA. The direct microspectrophotometric determination of the fraction of discarded DNA in elimination mitoses has never been reported. This method would probably yield very ambiguous results, because it is known that chromatin remaining in the cytoplasm will be rapidly degraded. All four quantitative DNA determination techniques used have their specific advantages and disadvantages. Feulgen-microspectrophotometric DNA determinations can be standardized, but the method has the drawback that the uptake of Feulgen dye seems to depend on a number of nonbiological parameters, because the determined DNA values change considerably, both during spermiogenesis and oogenesis (see, e.g., Pasternak and Barrell 1976). The same is probably also true for the variation in the stainability with the dye diphenylamine. However, this technique, together with isotope dilution, has the advantage of being a relatively fast and simple one. The disadvantage of the isotope dilution method resides in the fact that one is never sure whether the added, radioactively labeled DNA behaves in the same way during the extraction and purification procedure as the nuclear DNA, whose amount one wants to determine. Finally, DNA renaturation kinetics do not only allow to determine the genome size, but also the genomic complexity (Britten and Kohne 1968). The problem with this method is its rather low sensitivity, since small differences in genomic sizes cannot be detected.

7.2 Different Amounts of Eliminated DNA
in *Ascaris lumbricoides* var. *suum* and *Parascaris equorum*

Table 2 presents the sizes of genomes and the fractions of eliminated DNA for *Ascaris lumbricoides* and *Parascaris equorum*, as they have been reported by different laboratories. For *Ascaris lumbricoides*, the absolute values for the germ line and somatic genomes vary considerably between the five listed research groups. They range from 0.32 pg to 0.68 pg for the haploid germ line and from 0.145 pg to 0.46 pg for the haploid somatic genome. On the other hand, the calculated amounts of germ line-limited DNA sequences are in reasonable agreement; they range from 22 to 34%, apart from the value of 56% reported by Davis et al. (1979). From Table 2 it is also evident that the discrepancies between the presented sizes of genomes are less a function of different DNA determination

Table 2. Genome size and amount of eliminated DNA in *Ascaris lumbricoides* var. *suum* and *Parascaris equorum*

	Genome size (n)		Eliminated DNA		Method	Reference
	Germ line (pg)	Soma (pg)	(pg)	(%)		
Ascaris lumbricoides var. *suum*	0.63	0.46	0.17	27	Isotope dilution	Tobler et al. (1972)
	0.32	0.25	0.07	22	Feulgen-microspectrophotometry, Diphenylamine assay	Moritz and Roth (1976)
	0.68	0.45	0.23	34	Feulgen-microspectrophotometry	Pasternak and Barrell (1976)
	0.46	0.31	0.15	33	DNA renaturation kinetics	Goldstein and Straus (1978)
	0.33	0.145	0.185	56	Isotope dilution, Diphenylamine assay	Davis et al. (1979)
Parascaris equorum var. *univalens*	1.2–2.1	0.25	0.95–1.85	79–88	Feulgen-microspectrophotometry, Diphenylamine assay	Moritz and Roth (1976)
	ND[a]	ND	ND	~70	Estimated from the amount of heterochromatin in metaphase chromosomes	Goday and Pimpinelli (1984)
Parascaris equorum var. *bivalens*	ND	ND	ND	>90	Feulgen-microspectrophotometry of individual chromosomes	Moritz (1977)

[a] Not determined.

methods applied, but rather depend on the laboratory in which the experiments have been carried out. A comparison of the fraction of cast-off DNA in *Ascaris lumbricoides* and *Parascaris equorum* reveals that the amount of eliminated DNA is much larger in the latter species. Interestingly, it seems that the genome sizes in the somatic cell lines of the two species are about equal, but that the germ line genome of *Parascaris equorum* is about three to four times larger than that of *Ascaris lumbricoides*. According to Moritz (1970a, b, and personal communication), the value of 1.2–2.1 pg DNA per haploid germ line genome of *Parasaris equorum* var. *univalens* (see Table 2) signifies a real variation in the amount of germ line DNA between individuals of this species, and not just a variation which is due to technical difficulties.

In *Parascaris equorum* var. *univalens*, roughly 1 to 2 pg of nuclear DNA (about 80%) becomes eliminated from the presumptive somatic cells during the chromatin elimination process (cf. Table 2). This represents a substantial amount of DNA and hence potential genetic information. In *Ascaris lumbricoides* var. *suum*, about one fourth of the total nuclear DNA seems to be eliminated. If we take an amount of 0.17 pg cast-off DNA per haploid genome (Tobler et al. 1972), a value which is very close to the mean of all reported values for *Ascaris lumbricoides* (cf. Table 2), this amount of DNA equals 0.10×10^{12} daltons, 1.6×10^5 kb or approximately 54 mm of DNA in length. Even if this amount of DNA is about 6–11 times less than the quantity of DNA that seems to be eliminated in *Parascaris equorum* (cf. Table 2), it still represents a very large amount of DNA.

Noll and Bielka (1968) were the first to determine the relative amounts of DNA during early embryogenesis of *Ascaris lumbricoides* var. *suum*. They noted a decrease of DNA content per cell during development, and argued that the zygote contains a large amount of cytoplasmic DNA which is replicated at a slower pace than the chromosomal DNA. Searcy and MacInnis (1970), using DNA renaturation kinetics, determined the genome size of dissected ovaries, which contain somatic and germ line cells in a yet unknown proportion. They reported a MW of 1.5×10^{11}, which corresponds roughly to 0.25 pg DNA. Both results have not been included in Table 2, because the amount of eliminated DNA has not been measured in either case.

7.3 Comparison of the Quantity of Eliminated DNA in Different Taxonomic Groups

To my knowledge, the amount of eliminated DNA has only been determined in some copepods and ciliates, apart from the two nematode species I have just mentioned. According to S. Beermann (1977), using Feulgen-spectrophotometry, three different *Cyclops* species lose about 50% of their DNA from presomatic cells during the process of chromatin elimination (see Hennig this Vol., for more detailed information). In ciliates, as little as about 10% of the DNA sequences present in micronuclei seem to be missing in macronuclei in *Tetrahymena thermophila* and *T. pyriformis* (cf. Table 1). On the other hand, about 66% of the chromosomes in a first elimination cycle and more than 90% of the remaining DNA sequences during a second elimination cycle are expelled from the developing macronucleus in *Stylonychia mytilus*, thus leaving less than 3% of the original micronuclear DNA sequences in the macronucleus (see Ammermann 1986, and Steinbrück this Vol., for review). No quantitative DNA measurements before and after chromosome elimination are available from insects. However, cytological observations revealed that, e.g., in Cecidomyidae, somatic cells contain between 3 and 12 chromosomes, depending on the species and/or the sex. In the germ line cells, the chromosome number varies between about 20 and 80 chromosomes, again depending on the species and/or the sex of the individuals (see Matuszewski 1982, for review). Similarly, Bauer and W. Beermann (1952a, b) reported that in several Chironomidae species, the haploid number of somatic chromosomes is either 2 or 3. In the germ line cells, however, 1 to 26 additional chromosomes per haploid chromosome set were counted. The number of these germ line-limited chromosomes may vary within an individual, between individuals of the same species as well as between different species (Bauer and W. Beermann 1952b). Because most of the germ line-limited chromosomes are of about equal size as those present in somatic cells, one can roughly estimate the amount of chromosomal material which becomes expelled from the presumptive somatic cells during the chromosome elimination process. These figures vary between 25% (1 out of 4 chromosomes) and 93% (26 out of 28 chromosomes; cf. Table 1 in Bauer and W. Beermann 1952b).

8 Molecular Characterization of the Eliminated DNA Sequences

8.1 Density Gradient Analysis

Bielka et al. (1968) were the first who analyzed DNA of a chromatin eliminating nematode applying biochemical techniques. They isolated DNA from eggs and gastrulae of *Ascaris lumbricoides* and compared their buoyant density patterns by CsCl gradient centrifugation. According to their data, DNA preparations from eggs and gastrulae sediment in one main peak with a density of 1.697 g cm^{-3}. Egg DNA shows a second peak with a density at 1.685 g cm^{-3}; this fraction is considerably reduced in gastrula DNA. Bielka et al. (1968) assumed that this fraction might consist of mitochondrial DNA. Moreover, egg DNA contains a shoulder on the light side of the main peak DNA at a density of 1.693 g cm^{-3}, which becomes also substantially reduced in DNA from gastrula stages. It was suggested that this smaller DNA fraction represents satellite DNA.

In a study to specifically investigate the properties of the eliminated DNA sequences in *Ascaris lumbricoides*, we also compared the density profiles of germ line and somatic DNA in CsCl gradients (Tobler et al. 1972). DNA from spermatids and 4-cell stages were used as germ line DNA sequences, larval DNA served as a source for somatic DNA, since it is known that larval stages of *Ascaris lumbricoides* contain only few primordial germ cells. Essentially, the results of Bielka et al. (1968) have been confirmed, except that the DNA satellite at a density of 1.693 g cm^{-3} was not evident in our preparations of germ line and somatic DNA.

The results of Bielka et al. (1968) have been extended by demonstrating that the light peak DNA in 4-cell stages which corresponds to that of eggs consists indeed of mitochondrial DNA (Tobler et al. 1972, Tobler and Gut 1974). Furthermore, in our DNA preparations, the density patterns of spermatid and larval DNA were indistinguishable, so that we concluded that the eliminated DNA does not differ in its buoyant density from nuclear somatic and germ line DNA. Goldstein and Straus (1978) determined buoyant densities of 1.695 g cm^{-3} for somatic and of 1.697 g cm^{-3} for germ line DNA, which is in good agreement with the earlier results of Bielka et al. (1968) and Tobler et al. (1972).

Density gradient analyses of DNA from *Ascaris lumbricoides* before and after elimination have also been carried out by Moritz and Roth (1976), Roth (1979), and independently by Kilejian and MacInnis (1976). Their results are essentially in accord with each other; both groups report that germ line DNA contains a main band DNA at a density of 1.700 g cm^{-3} and a light satellite at 1.697 g cm^{-3} or 1.696 g cm^{-3}, not sharply separated from the peak of main band DNA. This satellite is absent in DNA isolated from larvae or from three different somatic tissues. Kilejian and MacInnis (1976) and Roth (1979) moreover report the presence of a major DNA fraction with a density of 1.690 g cm^{-3} in fertilized eggs and 4-cell stages, which is diminished in DNA from spermatids and even more so in DNA isolated from intestines. An additional minor heavy satellite with a density of 1.710 g cm^{-3} was present in egg and muscle DNA.

How should these seemingly conflicting reports from five different laboratories be reconciled? I think that if the density values of Moritz and Roth (1976), Roth (1979) and Kilejian and MacInnis (1976) are shifted by 0.003 g cm^{-3} units to equate the main band values at 1.697 g cm^{-3}, then their germ line satellite peak at 1.697 g cm^{-3} or 1.696 g cm^{-3} would coincide with the 1.693 g cm^{-3} band reported by Bielka et al. (1968), a suggestion already expressed by Kilejian and MacInnis (1976) to bring their results more in line with those of Bielka et al. (1968). The fact that this light satellite DNA has not been found by Tobler et al. (1972) and Goldstein and Straus (1978) may result from incomplete resolution. From recent DNA isolation, cloning, and sequencing studies we do know that this satellite fraction must be present in germ line DNA (see Sect. 8.4). The suggested shift by 0.003 g cm^{-3} units would furthermore allow uniting the different data on the light DNA satellite which consists of mitochondrial DNA. The corrected values for this organellar DNA would be 1.685 g cm^{-3} (Bielka et al. 1968), 1.686 g cm^{-3} (Tobler et al. 1972, Tobler and Gut 1974) and 1.687 g cm^{-3} (Carter et al. 1972, Kilejian and MacInnis 1976, Moritz and Roth 1978, Roth 1979). It remains to be explained why a minor heavy satellite with a density of 1.710 g cm^{-3} in egg and muscle DNA was evident only in the DNA preparations by Kilejian and MacInnis (1976), but was not found in DNA isolated from uterine wall, intestines, and spermatids by the same researchers. Personally I think that this heavy satellite might be a glycogen contaminant, since it is known that eggs and muscle tissues contain very large amounts of glycogen (Bueding and Orrell 1964, Kuhn and Tobler 1978). Even if the fraction with a density of 1.710 g cm^{-3} consisted of DNA, it can certainly not be involved in the chromatin elimination process, since this fraction seems to be present or absent both in germ line and somatic cells (Kilejian and MacInnis 1976).

8.2 DNA Renaturation Kinetics

In order to study the composition of the eliminated DNA sequences in *Ascaris lumbricoides*, we compared the reassociation kinetics of DNA isolated from germ line and somatic cells (Tobler et al. 1972). DNA extracted from spermatids and 4-cell stages was used as a source for germ line DNA sequences, DNA from larvae as a source for somatic DNA, respectively. According to our experimental results, we had to conclude that about 10% of the retained somatic DNA sequences are repetitious with an average family size of about 6000 copies, the rest appears to consist of unique sequences. However, germ line DNA from spermatids contains about 23% fast-renaturing DNA sequences repeated about 7000–10,000 times in the germ line genome. Because about 27% of germ line DNA is expelled from presumptive somatic nuclei during the chromatin elimination process, we further concluded that the eliminated DNA consists of repetitive and unique sequences in a ratio of approximately 1 : 1 (Tobler et al. 1972). In 4-cell stages, the fraction of highly repetitious DNA sequences is increased by about 25% as compared to spermatid DNA. Density gradient centrifugations and electron microscopy revealed that 4-cell stages of *Ascaris lumbricoides* contain about 25–40% mitochondrial DNA (Tobler et al. 1972, Tobler and Gut 1974). Because in our preparations DNA has been isolated from whole 4-cell stages and not from isolated nuclei, the presence of a major fraction of mitochondrial DNA was to be expected in these DNA reassociation experiments.

Moritz and Roth (1976), using slightly different techniques and methodology, came to the conclusion that only highly repetitious DNA sequences of the satellite type are discarded during the chromatin elimination process in *Ascaris lumbricoides* and *Parascaris equorum*. In an attempt to resolve this conflicting situation, Goldstein and Straus (1978) repeated the reassociation kinetic experiments and reported that the eliminated DNA sequences in *Ascaris lumbricoides* consist of repetitive and unique sequences in a ratio of approximately 1 : 1, thus confirming our earlier findings and conclusions. There is the possibility that the differences between the results of Moritz's group (Moritz and Roth 1976, Roth 1979, Roth and Moritz 1981) on the one hand and Goldstein and Straus (1978) and ours (Tobler et al. 1972) on the other, may have arisen from real differences between presumed subspecies of *Ascaris lumbricoides*, as has been suggested by Goldstein and Straus (1978). The *Ascaris* worms we had used in 1972 and those of Goldstein and Straus are of North American origin, while those of Moritz and Roth are from Europe. Goldstein and Straus (1978) argued that there is some evidence for karyotype differences between these two presumed subspecies, the haploid number of the North American variety being $n = 12$ and that of the European variety amounting to $n = 24$. However, for reasons I have outlined in Section 5.3, I do not believe in actual differences in chromosome number between the two assumed subspecies. At the time our original DNA renaturation experiments with germ line and somatic DNA of *Ascaris lumbricoides* were carried out, foldback DNA was not yet known; therefore this DNA fraction is included in the portion of repetitious DNA sequences determined by us. We have since repeated the DNA renaturation experiments with germ line and somatic DNA of *Ascaris lumbricoides* a number of times in our laboratory, and always found between 2 and

9% repetitive DNA sequences in the somatic and 16 to 25% in the germ line cells, depending on the experimental methods applied and whether or not foldback DNA was subtracted. However, in all cases a clearly measurable fraction of repetitive DNA sequences was also present in the genome of the somatic cells. That somatic DNA must contain a fraction of repetitive DNA sequences was furthermore confirmed by saturation hybridization experiments with satellite DNA (see above) and by determining the interspersion pattern of repetitive and unique DNA sequences (see Sect. 8.6). In summary, the eliminated portion of the genome in *Ascaris lumbricoides* does certainly contain a large portion of highly repetitive DNA sequences, but also nonrepetitive DNA sequences (for further evidence see Sect. 8.5). However, I no longer believe that the eliminated genome consists of repetitive and unique DNA sequences in a ratio of 1 : 1, but rather that a larger portion of repetitive sequences becomes eliminated from the presumptive somatic genome.

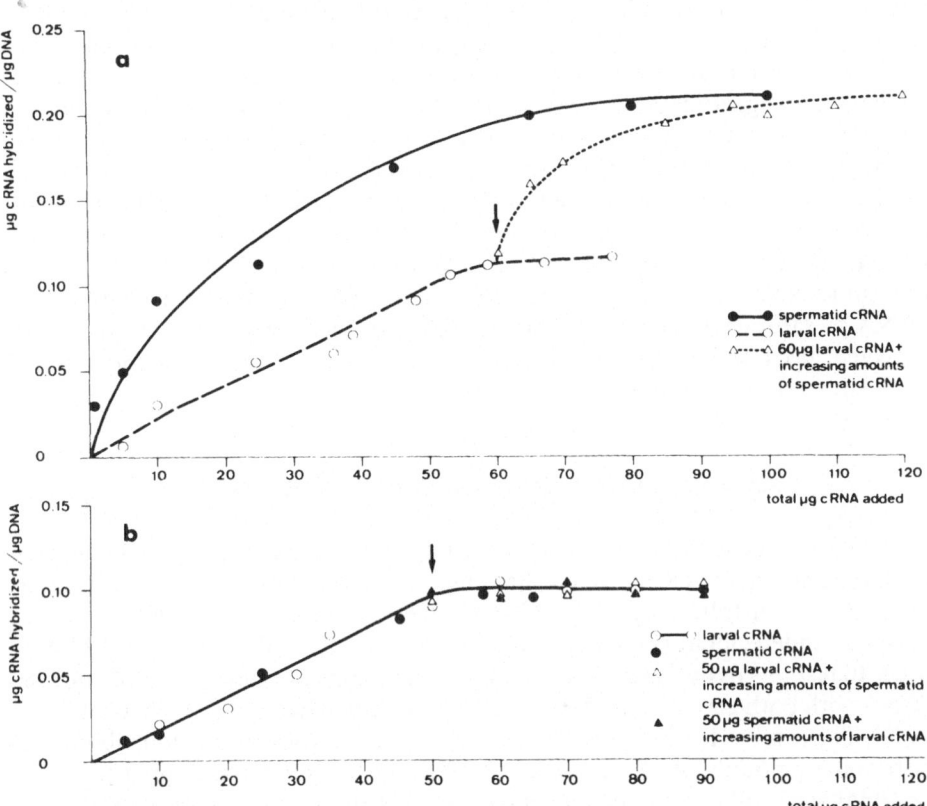

Fig. 11. a Saturation of spermatid DNA isolated from *Ascaris lumbricoides* with increasing amounts of larval (o--o) and spermatid (●—●) cRNA. After the saturation value for larval cRNA was determined (60 μg/larval cRNA), increasing amounts of spermatid cRNA (△·····△) were hybridized in the presence of saturating amounts of larval cRNA. **b** Saturation of *Ascaris* larval DNA with increasing amounts of larval (o—o) and spermatid (●) cRNA. In additional experiments, saturating amounts of larval cRNA in the presence of increasing amounts of spermatid cRNA (△) or saturating amounts of spermatid cRNA with increasing amounts of larval cRNA (▲) were hybridized to larval DNA. (Tobler et al. 1972)

Because germ line cells of *Ascaris lumbricoides* clearly contain more repetitive DNA sequences than somatic cells, the important question arose whether the germ line-limited repetitive DNA sequences differ in a qualitative or merely in a quantitative way from the repetitive DNA sequences retained in the somatic cell lineage. To examine this problem, sequential hybridization experiments were performed (Tobler et al. 1972, cf. Fig. 11). At the time these experiments were done, radioactively labeled DNA from *Ascaris lumbricoides* was not readily available. We therefore used complementary RNA (cRNA) derived from spermatid (germ line) and larval (soma) DNA in our hybridization experiments. The two cRNA's were hybridized in increasing amounts to either spermatic or larval DNA. As Fig. 11 a demonstrates, larval cRNA at saturation anneals to about 11% of the spermatid DNA, and spermatid cRNA to about twice as much. If saturating amounts of larval cRNA are hybridized together with increasing amounts of spermatid cRNA, the extent of hybridization rises again to reach the same saturation level as with spermatid cRNA alone. This result suggests that germ line DNA does contain sequences which are no longer present in larval DNA. Because filter hybridizations and low Cot conditions were used, only the repetitious sequences of the *Ascaris* genome have been analyzed in these experiments. Therefore, the above experiment allows only the conclusion that the eliminated repetitious sequences are distinct from the repetitious sequences retained in somatic cells. The same conclusion has been reached by Roth (1979). In control experiments, larval or spermatid cRNA was annealed to larval DNA (Fig. 11 b). As was to be expected, the saturation plateaus are identical for both cRNA's. Moreover, the saturation level is not changed by the addition of increasing amounts of spermatid cRNA to saturating amounts of larval cRNA or vice versa. This result is to be expected, because all somatic DNA sequences must, of course, be contained in the genome of the germ line cells.

8.3 Elimination of rRNA and/or tRNA Coding Genes?

It has been proposed that chromatin diminution serves the purpose of discarding large amounts of rRNA genes that might have selectively been amplified during oogenesis or following fertilization (Kaulenas and Fairbairn 1968), or which persist as independent rDNA episomes in the germ line (Wallace et al. 1971). In order to test this hypothesis, saturation hybridization experiments of 18S and 28S rRNA with both germ line and somatic DNA were carried out (Tobler et al. 1972, 1974). The results clearly show that germ line and somatic cells contain about proportionate amounts of rRNA genes and that no quantitative elimination of ribosomal genes takes place during chromatin elimination in *Ascaris lumbricoides*. Recently, we have analyzed the structural organization of the ribosomal genes in *Ascaris lumbricoides* in much more detail (Back et al. 1984 a, b, c) and found that the rDNA cluster of *Ascaris lumbricoides* is located on a single autosomal position, as could be demonstrated by in situ hybridization. Moreover, the rDNA cluster consists of two main classes of rRNA coding genes, 8.8 kb and 8.4 kb in length, the heterogeneity being the result of a 450-bp-long insertion located in the nontranscribed spacer region of the larger rDNA unit (Back et al. 1984a). The quantitative ratio of the two rDNA size classes is on the average roughly 10:1 in the investigated wild-type population of *Ascaris lumbricoides*, but varies to a great extent between different individuals (Back et al. 1984b). However, since no differences have been detected in the hybridization pattern of the two rDNA's between germ line and somatic cells in any single individual tested, one can further conclude that chromatin diminution does not qualitatively change the rDNA pattern of germ line versus somatic cells (Back et al. 1984b). Using saturation hybridization of radioactively labeled tRNA with germ line and somatic DNA of *Ascaris lumbricoides*, we were also able to show that the number of tRNA coding genes does not vary in a measurable way before and after chromatin elimination (Aeby 1979). To summarize, neither rRNA nor tRNA coding genes seem to be present in the eliminated chromatin of *Ascaris lumbricoides*, at least not in a measurable amount.

8.4 Elimination of Satellite DNA Sequences

In 1976, Moritz and Roth reported the presence of a highly repetitive DNA fraction with satellite DNA properties in *Ascaris lumbricoides* and *Parascaris equorum*. They later on succeeded in preparatively isolating an AT-rich DNA satellite from germ line cells of *Ascaris lumbricoides* (Roth and Moritz 1981) and characterized the DNA satellite by restriction enzyme analysis. They also reported that the germ line-limited satellite DNA is composed entirely of two related families of repeated sequences, one repeating unit being 125 bp, the other one 131 bp long. The germ line contains about 5×10^5 copies of repeating units, but a limited number of copies is also retained in the somatic cells (Roth and Moritz 1981). Sequencing studies (Streeck et al. 1982) revealed that the originally communicated length values of the prototype sequences had been overestimated, and demonstrated that both AT-rich variants are actually only 123 bp long. Furthermore, they have shown that the two prototype sequences differ in about 20% of their base sequence (Streeck et al. 1982).

Independently of Moritz's group, we have carried out a similar analysis on the structural organization of the satellite DNA contained in the eliminated genome of *Ascaris lumbricoides*, thus confirming and extending several of their findings (Müller et al. 1982a, b). Because they used uncloned material for their experiments, the reported prototype sequence is representative for the major portion of the satellite DNA. However, this method does not allow demonstration of the diversity between and within different variant classes. We have isolated, cloned, and sequenced several restriction endonuclease fragments derived from the germ line DNA satellite (Müller et al. 1982a, b). A comparison of all the determined sequences, which differ only by small deletions, insertions, and single base substitutions, allowed us to establish a consensus sequence (Müller et al. 1982b, cf. Fig. 12). This sequence is in good agreement with those published by Streck et al. (1982). One of their prototype sequences is very similar to, but not identical with, our consensus sequence, the other one can be arranged among our Bam fragments (Müller et al. 1982b). The comparative analysis of our 121 bp long consensus sequence has provided much evidence for an internal short range periodicity

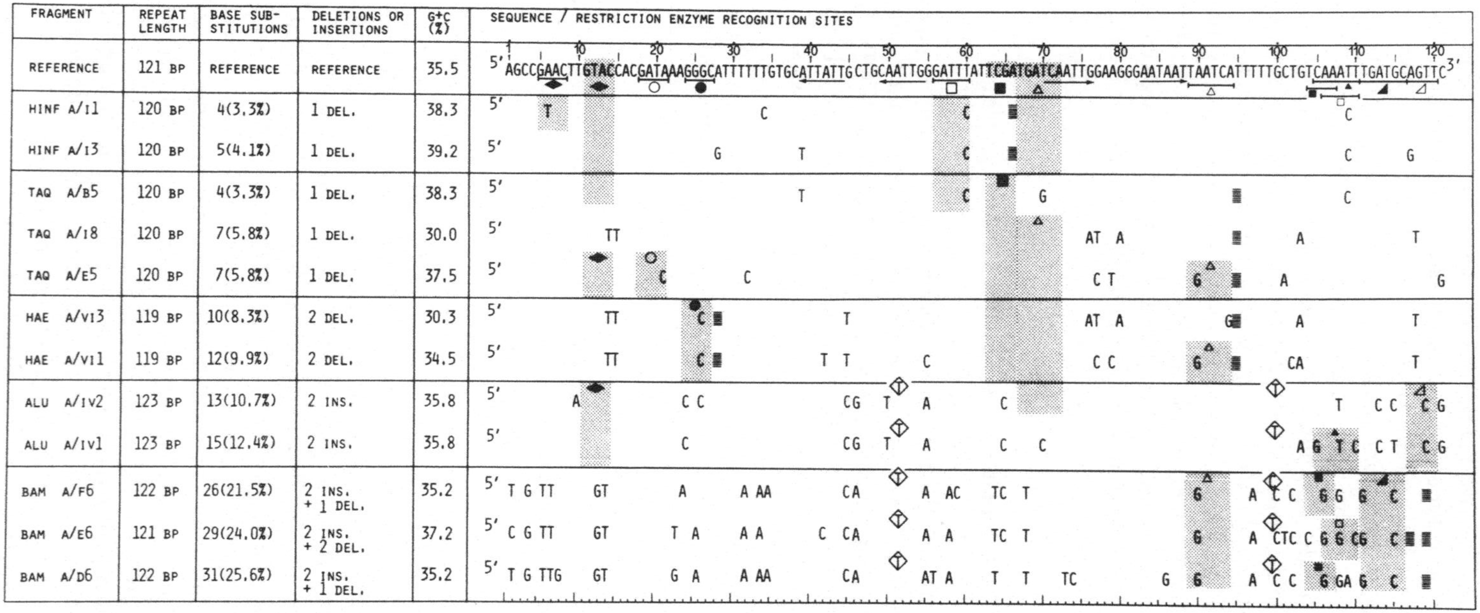

Fig. 12. Consensus sequence (*on top*) and sequences of 12 individual clones of highly repetitive germ line DNA of *Ascaris lumbricoides*. The reference sequence represents the consensus sequence deduced from all sequenced DNA fragments. The different sequences are listed in order of decreasing homology. Base substitutions relative to the reference sequence are indicated by the respective bases, deletions by ▬ and insertions by ◇. *Arrows underlining the reference sequence designate small palindromic sequences.* The restriction enzyme recognition sites are ◆ Rsa I, ○ Mbo I, ● Hae III, □ Hinf I, ▬ Taq I, △ Bcl I, ▲ Eco RI, ▲ Bam HI, △ Alu I; they are indicated by *stippled areas.* (Müller et al. 1982b)

of 11 bp in length with the deduced sequence 5'GCA(^T_A)TT(^T_G)TGAT (Müller et al. 1982b). It is therefore likely that the *Ascaris* satellite has evolved from an ancestral variant of this 11 bp long prototype sequence.

In order to determine the amount of eliminated satellite DNA during chromatin diminution, saturation hybridization experiments of germ line and somatic DNA with labeled monomeric satellite fragments were performed (Müller et al. 1982a, b). The difference in the two saturation values clearly demonstrates that chromatin diminution removes over 99.5% of the satellite DNA sequences from the presumptive somatic cells. Because we estimate the satellite to represent about 20% of the germ line genome, corresponding to about 10^6 copies of the 121 bp

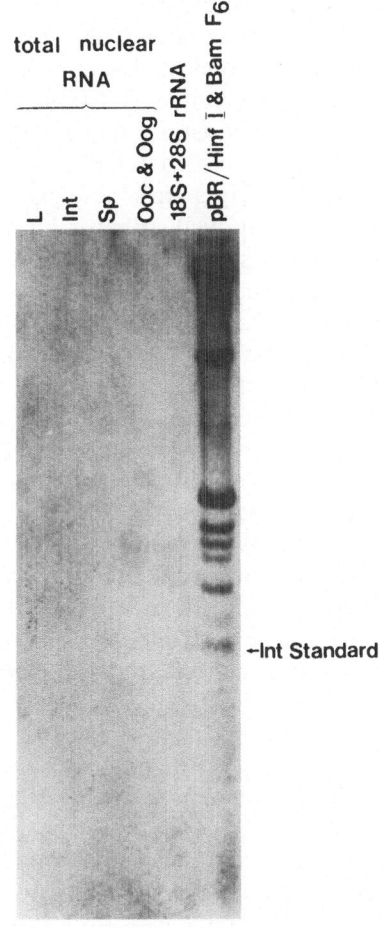

Fig. 13. Total nuclear RNA's were isolated from larvae (*L*), intestines (*Int*), spermatids (*Sp*), oocytes (*Ooc*) and oogonies (*Oog*) of *Ascaris lumbricoides*. Five µg aliquots of these different RNA's, as well as 0.5 µg of 18S and 28S rRNA, 50 ng of pBR 322/Hinf I together with 50 ng of a cloned satellite monomer (corresponding roughly to 1.25 ng of the cloned monomer) as an internal standard were run on a 2% agarose slab gel, transferred to DBM paper and hybridized with labeled cloned satellite monomers. Hybridization of the labeled probe is only detectable with the internal standard. (Müller et al. 1982a)

repeating units (Müller et al. 1982 b), somatic DNA would at most contain roughly 5000 copies. From the fact that development and differentiation of the somatic cell lines in *Ascaris lumbricoides* proceed normally after chromatin diminution in the presomatic cells has taken place, one has to conclude that the germ line-limited DNA sequences are not required for the normal function of somatic cells and tissues. It is interesting that a similar situation seems to exist in DNA-eliminating ciliates. The generative micronuclei contain a much larger amount of repetitive DNA sequences than the vegetative macronuclei (see Steinbrück this Vol.). Because micronuclei correspond to germ line, and macronuclei to somatic nuclei of multicellular organisms, the elimination process seems to cast off predominantly (but not exclusively, see Sect. 8.5) repetitive DNA sequences from the somatic cell lineages in both instances.

The question of whether or not satellite DNA sequences are transcribed in different germ line and somatic tissues of *Ascaris lumbricoides* has been investigated by nucleic acid hybridization experiments (Müller et al. 1982 a, b; cf. Fig. 13). Our results, depicted in Fig. 13, demonstrate that satellite DNA sequences of *Ascaris lumbricoides* are either not transcribed in larvae, intestines, spermatids, oogonies, and oocytes, or that the transcripts are so few that they have gone undetected by the experimental procedures applied. What is known about the function of this *Ascaris* satellite DNA, which is carried in the germ line cells but becomes mostly eliminated from the somatic cells? As is the case for all DNA satellites of eukaryotes described so far, nothing definite is known about its function (see, e.g., Bostock 1980, Singer 1982, Lewin 1982). However, our data are consistent with the notion that satellite DNA exerts its biological effects in processes that are inherent to the germ line cells. Therefore, if this satellite DNA has any function at all, it must be germ line-limited. Otherwise one would have to assume that the same function could also be carried out by the few satellite DNA copies remaining in the somatic cells, which is rather unlikely.

Recently, Dawson et al. (1984) reported that in the ciliate *Oxytricha fallax*, highly repetitious DNA sequences present in the generative micronuclei are eliminated from the vegetative macronuclei. They also concluded that if these tandemly repeated sequences have any function at all, it must be limited to the germ line and that these DNA sequences are probably not required for vegetative growth.

8.5 Elimination of Nonsatellite DNA Sequences

The analysis of putative, eliminated, single copy DNA sequences or sequences with a low redundancy in *Ascaris lumbricoides* proved to be difficult for the following two reasons: (1) Such sequences are relatively rare in the genome and therefore not easy to isolate, (2) a specific DNA probe is needed to test for its presence in the germ line and absence in the somatic genome. In order to overcome these problems, a gene library of *Ascaris lumbricoides* germ line DNA was established in our laboratory (Aeby et al. in preparation). This gene library enabled us to pick out several clones which contain eliminated, nonsatellite DNA sequences. For instance, one of them gives rise to an Eco RI–Hind III fragment

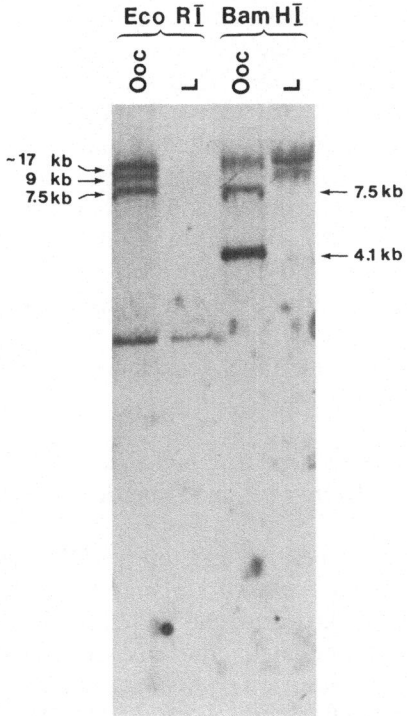

Fig. 14. Hybridization pattern of Eco RI and Bam HI digested total DNA from oocytes (*Ooc*) and larvae (*L*) of *Ascaris lumbricoides*. The restriction fragments were separated on an agarose slab gel, Southern transferred and hybridized with a labeled subcloned DNA fragment originally arising from the *Ascaris* germ line gene library. For the details of the preparation of the labeled hybridization probe see text. The three long DNA fragments of about 17, 9, and 7.5 kb, clearly present in Eco RI digests of oocyte DNA, are absent in digests of larval DNA, whereas the smaller fourth DNA fragment is weaker in the larval than in the oocyte DNA digest. In Bam HI digests, the two 7.5- and 4.1-kb-long DNA fragments are present in the digests of oocyte DNA, but missing in those of larval DNA. (Aeby et al. in preparation)

of about 1050 bp, which hybridizes clearly to a Southern transfer of Eco RI or Bam HI digested germ line and somatic DNA (see Fig. 14). However, whereas this DNA probe hybridizes strongly to DNA fragments of about 17, 9, 7.5 kb and a much smaller band in Eco RI digested DNA derived from oocytes, there is only a faint hybridization signal corresponding to the small band in digests of larval DNA. Similarly, in Bam HI digests, a strong band at 4.1 kb and a weaker one at 7.5 kb are visible in digests of oocytes, but absent in those of larval DNA. These results therefore demonstrate quite convincingly that DNA sequences other than satellite DNA must be contained in the germ line-limited chromatin and therefore be expelled from the presumptive somatic cells during the process of chromatin elimination (Aeby et al. in preparation). So far we do not know whether or not this particular DNA sequence, or others we have isolated, exerts any function in the germ cell lineage. However, experiments to test whether some of these DNA sequences are transcribed in a stage-specific way in the germ line cell lineage are currently under way in our laboratory.

Earlier autoradiographic studies carried out in *Parascaris equorum* by Moritz (1970 b) indicated that the germ line-limited chromatin may be transcriptionally active in gametogonial stages; but its activity seems to be very low, if there is any, during oocyte maturation and early cleavage stages. In the cecidomyid species *Wachtliella persicariae*, the germ line-limited chromosomes decondense and synthesize RNA during oogenesis, in contrast to the compact somatic chromosomes, which are inactivated during the first period of oocyte growth (Kunz et al. 1970, Kunz 1970). There is also autoradiographic evidence that the micronucleus in three species of *Paramecium* and *Tetrahymena* is transcriptionally active (see Murti and Prescott 1970, for early survey and Steinbrück this Vol., for recent review). All these autoradiographic data support, of course, the hypothesis of a germ line-specific function of these DNA sequences, but obviously do not prove it.

8.6 Interspersion of Repetitive and Unique DNA Sequences in the Germ Line and Somatic Genome

To study the pattern of repetitive and unique DNA sequences in the eliminated genome of *Ascaris lumbricoides* var. *suum*, we have determined the arrangement of these sequences in the germ line and somatic genomes (Landolt and Tobler 1980), and found that the somatic DNA of *Ascaris lumbricoides* is arranged in a short-period interspersion or *Xenopus* pattern, as is the case for most other eukaryotes (see, e.g., Jelinek and Schmid 1982). Following the protocol of Davidson et al. (1973), we calculated that the somatic genome of *Ascaris lumbricoides* consists of about 1% clustered, repetitive sequences, about 12% single copy DNA sequences in short-period interspersion (530 bp) and at least 22% single copy DNA sequences in long-period interspersion (5100 bp). These experiments were repeated and extended to an analysis of the germ line genome taken from spermatids (Landolt and Tobler in preparation). According to our most recent calculations, the germ line-limited genome which becomes discarded from the presumptive somatic cells during chromatin elimination consists of about 60–75% highly respective DNA sequences, mostly or exclusively of the satellite type, the rest belonging to the low abundant or single copy DNA fractions. This portion seems to arise from the genuine fraction of single copy DNA, uninterrupted by moderately repetitive DNA sequences (Landolt and Tobler in preparation). Of course, all these calculations depend heavily on the determined quantity of eliminated DNA. The values given above correspond to an amount of 27% eliminated chromatin (Tobler et al. 1972); if less chromatin is discarded, then the number of single copy DNA sequences decreases accordingly, and vice versa.

Roth (1979) concluded from renaturation kinetics of long somatic DNA fragments that the DNA sequence arrangement of *Ascaris lumbricoides* does not follow the short-period but rather the long-period interspersion or *Drosophila* pattern. However, the approach he used is not very sensitive to detect interspersion of short repetitive DNA elements (see, e.g., Emmons et al. 1980). Schachat et al. (1978) and Beauchamp et al. (1979) also reported a long-period interspersion pattern in the genomes of the free-living nematode species *Caenorhabditis elegans* and *Panagrellus silusiae*, respectively. But again, the same caution should be expressed as above, because Emmons et al. (1980), combining reassociation kinetics with electron microscopic techniques, have shown that the genome of *Caenorhabditis elegans* is arranged in the characteristic short-period interspersion pattern shared by most eukaryotes.

9 The Mechanism of Chromatin Elimination

9.1 Cytological and Electron Microscopical Observations

Very little is known about the mechanism of chromosome and chromatin elimination, and recent cytological data on this process are extremely scarce. As we have seen in Sections 5 and 6, there is experimental evidence that cytoplasmic factors are responsible for the induction or prevention of chromosome or chromatin elimination. The elimination process may involve whole chromosomes or only parts of chromosomes (cf. Sect. 5.1); in the latter case we are faced with the additional problem that the chromosomes must break up into pieces, some of which will be discarded during the elimination process. In certain species, as in *Parascaris equorum* var. *univalens* (see, e.g., Goday and Pimpinelli 1984), or in *Cyclops divulsus* (S. Beermann 1977), light microscopic evidence indicates that only telomeric heterochromatic blocks become eliminated in presomatic cells. On the other hand, other species like *Parascaris equorum* var. *bivalens* show, in addition to the telomeric heterochromatin, several small blocks of intercalary heterochromatin, which are also discarded during elimination (Goday and Pimpinelli 1984). In *Cyclops strenuus*, small heterochromatic segments destined to become eliminated in presomatic nuclei are scattered all along the chromosomes, whereas in *Cyclops furcifer*, heterochromatin is located at both ends, as well as in the centromere region (S. Beermann 1977). After elimination of intercalary chromatin, the chromosomal integrity is restored, hence controlled rejoining of the retained chromosomal fragments is required. In this context it is interesting that S. Beermann and Meyer (1980) presented electron microscopic evidence that in *Cyclops furcifer*, chromatin is eliminated in the form of numerous rings of variable sizes. Such chromatin rings have also been found during early diminution stages of *Cyclops divulsus* (S. Beermann 1984), a species that carries only terminal heterochromatic segments. Employing the Miller spreading techniques, S. Beermann (1984) furthermore detected nucleosomal chromatin rings during the critical stages of diminution, though in lower frequencies. Thus it remains uncertain whether all the eliminated DNA arises as circular molecules in both *Cyclops* species studied. Chromatin ring formation has also been discovered during macronucleus formation in *Stylonychia mytilus*, at a stage where chromatin elimination is known to occur (G. F. Meyer and Lipps 1980). It is therefore suggestive that elimination of DNA in the above-mentioned cases might involve a mechanism which is analogous to the excision of DNA in prokaryotes remove (see Hennig this Vol., for a more detailed discussion).

Experimental and descriptive studies of chromosome elimination in the cecidomyid fly *Miastor* sp. by Nicklas (1959) indicated that eliminated chromosomes possess morphologically normal chromosomal fibers during anaphase and that interlocking or sticking of daughter chromatid ends is not responsible for chromosome elimination. A comparative ultrastructural analysis of normal and elimination mitoses in *Heteropeza pygmaea* revealed that during anaphase, the disassembling microtubules which are connected with normally segregating chromosomes become increasingly coated with electron-opaque material (Fux 1974). In contrast, microtubules of eliminated chromosomes blocked in the interzone persist throughout anaphase and remain uncoated. Therefore, it is possible that in Cecidomyidae which undergo chromosome elimination, the microtubule coating material plays an important role in whether or not chromosomes are eliminated.

To my knowledge, only one report has been published on the ultrastructure of the shed ribonucleoprotein (originally termed elimination chromatin, cf. Sect. 5.1) in meiotic stages of oocytes in a lepidopteran species (Sorsa and Suomalainen 1975). According to their electron microscopic analysis, the discarded ribonucleoprotein is far less electron-dense than the chromosomal chromatin and shows a fuzzy fibrillar structure, the fibers having a diameter of about 100 Å.

9.2 Molecular Approach to Analyze the Process of Chromatin Diminution

We are completely ignorant of the molecular mechanism which leads to chromosome fragmentation and elimination. In a recent short note, Moritz (1984) proposed a model to explain the process of chromatin elimination in molecular-genetic terms. However, I feel that it is still premature to speculate about the mechanism if almost no experimental facts at the molecular level are available. To analyze the chromatin diminution process at the molecular level, we have screened our gene library of *Ascaris lumbricoides* germ line DNA (cf. Sect. 8.5) for clones which comprise satellite and nonsatellite DNA sequences. Such clones are good candidates for containing the putative elimination site. One of these clones has been analyzed in detail and partially sequenced (Aeby et al. in preparation). Interestingly, we found, just adjacent to the satellite fraction, a DNA sequence which has all the structural characteristics of a transposable element. The same transposon-like sequence is present in about 50 copies elsewhere in the *Ascaris* genome. Several other genomic clones containing this transposon-like element have been isolated and characterized. They seem to be inserted at many different sites in the *Ascaris* genome, although their structure is highly conserved. In all cases investigated thus far, the transposon-like element seems to be flanked on one side by sequences which are either eliminated or rearranged during the elimination process (Aeby et al. in preparation). These results led us to assume that this transposable element may somehow be involved in the chromatin elimination process.

As was already noted by B. McClintock a long time ago, displacements of transposable elements may generate chromosomal rearrangements or mutations (see Fedoroff 1984, for a recent review). Transposable elements have also been isolated and characterized in the free-living nematode *Caenorhabditis elegans* in the laboratories of S. W. Emmons and D. Hirsh (Emmons et al. 1983, Emmons and Yesner 1984, Ruan and Emmons 1984, Liao et al. 1983, Rosenzweig et al. 1983 a, b). They have shown that the *Caenorhabditis elegans* Bristol strain contains about 25–30, the Bergerac strain several hundred copies of the transposable element termed Tc1. Emmons and Yesner (1984) have recently demonstrated that the Tc1 transposable element undergoes excision at high frequency, but that this excision occurs primarily or entirely in the somatic tissues, whereas the Tc1 element is rather stably maintained in the germ line cells. Thus the excision of this transposon in *Caenorhabditis elegans* seems to be under the control of tissue-specific factors (Emmons and Yesner 1984). These experiments, therefore, reveal that transposable elements in *Caenorhabditis elegans* lead to differences in the arrangement of DNA sequences in the germ line and somatic genomes. However, I believe that it is still an open question whether or not the two processes of chromatin diminution and Tc1 excision are indeed related phenomena.

10 Chromosome and Chromatin Elimination Compared to Other Phenomena that Lead to Genome Alterations

10.1 The Dogma of Genetic Identity of Different Cell Types

There is an old dogma in biology which states that all cells of a higher organism contain identical sets of genetic information in their nuclei (see, e.g., Tobler 1972). Support for this so-called DNA constancy rule stems from the following facts and observations: (1) All cells of a multicellular organism are descendants of a single cell, the zygote, whereby the process of mitosis leads to equal assignment of the chromosomes and their genes to the daughter cells. (2) Direct measurements of the DNA content per diploid nucleus of various somatic cell types of an organism revealed equal amounts of DNA (see Vendrely 1955). The only exceptions are provided by polyploid and polytene cells. These exceptional cases of increased DNA content have often been interpreted as resulting from several additional rounds of replications of the entire genome. Provided that this assumption is correct (see, however, Sect. 10.2), such cells would merely differ in quantity but not in quality from the DNA content of diploid somatic cells. (3) Polytene chromosomes from different tissues of an individual show a nearly identical banding pattern (W. Beermann 1952). Because it has been possible, at least in some instances, to correlate a given gene with a specific chromosomal band (see W. Beermann 1972, Judd et al. 1972), it may be concluded that bands of polytene chromosomes represent genetic loci, although the simple correlation that one band equals one complementation group is certainly an overstatement (Lima-de-Faria 1975, Gall 1981). However, it is generally still accepted that the equal number and almost identical pattern of polytene chromosome bands in different tissues such as larval salivary glands, Malpighian tubules or fat body reflects the same number and sequence of genes in these tissues. (4) Early DNA renaturation studies by McCarthy and Hoyer (1964) indicated that DNA sequences derived from different tissues or from different stages of embryogenesis of the mouse are qualitatively and quantitatively identical. But with the discovery that the genome of all higher organisms contains repeated as well as unique DNA sequences (Britten and Kohne 1966, 1968), it became clear that the fraction of DNA which had been analyzed by the technique of McCarthy and Hoyer (1964) consisted exclusively of repeated and not of single copy DNA sequences. Therefore, genetic identity of cells from different mouse tissues was only shown for the reiterated portion of the genome. However, with the advent of molecular cloning techniques, it became possible to test whether or not genomic DNA sequences, isolated from the single copy or low redundancy range, are amplified, lost, rearranged, or modified in a tissue-specific and stage-specific manner. In general, such experiments revealed no differences in the arrangement or structure of the tested genes in different tissues of an organism. For instance, such an analysis was carried out with six developmentally regulated regions of genomic DNA sequences in *Drosophila*, five of which contained structural genes and the sixth region included a dispersed segment that mapped at about 30 different sites and seemed to be transposable (Levine et al. 1981). Their results revealed no indications of selective amplification, loss, modification or rearrangement of the six, analyzed, genomic DNA sequences in any of the tissues or developmental stages tested. (5) Experiments in developmental biology yielded so far the best evidence for the qualitative identity of the genetic material among different cell types of an organism. In several plant species it was possible to demonstrate that a single cell, which stems from differentiated adult plant tissue, may give rise to an entire plant (Steward et al. 1964). Even clearly differentiated single protoplasts, prepared from different adult tissues from a variety of plants, are able to regenerate and therefore reproduce asexually whole normal and fully fertile plants (see, e.g., Dale 1983, Davey 1983). Such a remarkable regenerative potential has never been demonstrated for entire animal cells. However, nuclear transplantation experiments in amphibians led to similar conclusions with respect to the genetic potential or differentiated cells. Nuclei from differentiated intestinal cells of *Xenopus* tadpoles which have been injected into enucleated eggs can promote the development of a normal and fertile frog (Gurdon and Uehlinger 1966), whereas enucleated eggs injected with nuclei from adult skin, lung, heart, and kidney tissues may develop into tadpoles (Laskey and Gurdon 1970, Gurdon et al. 1975). Why such nuclear transplants so far never developed into mature fertile frogs is not at all clear and open to different interpretations (cf. Gurdon 1974, DiBerardino et al. 1984). However, nuclear transplantation experiments in *Xenopus laevis* do show that the nucleus of the differentiated cell still contains the genetic information required to promote the development of tadpoles and that, therefore, cell differentiation does not depend on gene segregation, gene loss, or irreversible gene repression and/or alteration. (6) Similar conclusions can be drawn from some regeneration experiments and from the processes of transdifferentiation and transdetermination. In all these instances it has been demonstrated that the differentiated or determined state of cells may be altered under certain experimental conditions. In some salamanders, after complete removal of the lens, a new lens regenerates from the fully differentiated cells of the pigmented dorsal iris (see Yamada and McDevitt 1984). Such a change in the differentiated state (transdifferentiation) can also be provoked in cells cultured in vitro (see Okada 1983). Pluripotent transdifferentiation has furthermore been observed to occur in the anthomedusa *Podocoryne carnea*, in that striated muscle cells may transform into a great number of different cell types (Schmid and Alder 1984). The discovery of transdetermination in *Drosophila* provides another example of a change in cell heredity (Hadorn 1965). Cells of imaginal disks kept in culture in abdomens of adult flies are able to switch from one determined state into another, and to enter new developmental pathways (see, e.g., Hadorn 1978).

None of the above-mentioned examples of the reversibility of the determined or differentiated state proves that the genetic material of the cells remained unchanged during the process of cellular differentiation; they merely show that the genome has not become irreversibly altered, or that the genes in question which are necessary for the differentiation of the new cell type have not been eliminated. The same is true for the experimentally demonstrated developmental totipotency of differentiated cells in some plants and for the pluripotency of transplanted somatic nuclei into enucleated eggs of *Xenopus laevis* (see above). One should also keep in mind that the degree of differentiation attained in plants is generally much lower than that in higher animals. Moreover, totipotency of differentiated cells in animals has so far only been demonstrated for *Xenopus laevis* (see Gurdon 1974). Similar experiments carried out in related amphibian species have so far failed to be successful. Therefore, I think it is still premature to extrapolate the results from *Xenopus* to all other animal species. Furthermore, the DNA constancy rule is by no means universally accepted, and we will discuss the many exceptions to this rule in the following paragraph.

10.2 Cell Type-Specific Differences in DNA Content

Besides the obvious differences in the amount of DNA content between different cell types produced by chromosome or chromatin elimination (see Sects. 5 and 7), there are also some cases known in which terminally differentiated cells lose the whole set of their nuclear genetic information. Such a complete loss of nuclei has been noted to occur in mammalian erythrocytes, eye lens fiber cells in vertebrates, or sieve elements in plants. Of course, all these cell types are unable to undergo transcription or further cell division. Endopolyploidy, a phenomenon rather widespread both in the plant and animal kingdom (see Nagl 1978, for review), leads to a multiplication of the entire chromosomal set of a cell. This process may occur in a tissue-specific manner, as, e.g., in human liver cells. Polytenization results in lateral multiplication of chromatids without separation of the daughter chromatids, thus the number of chromosomes remains identical with those of cells not undergoing polytenization. Both processes may lead to amplification or underreplication of specific gene sequences (discussed extensively by Nagl 1976 and 1978). In this context it is interesting to note that during polytenization of *Drosophila* salivary gland chromosomes, the genes coding for ribosomal RNA which are located in the nonreplicating or severely underreplicated heterochromatic portion of the X and Y chromosomes are replicated, though to a lower degree than the euchromatic DNA portion (Hennig and Meer 1971, Spear and Gall 1973, Endow and Gall 1975). Because in these instances the endomitotic and polytenization cycles do not lead to equal replication of all genes, the gene balance must also become altered.

Amplification of specific genes represents another well-documented example for the violation of the DNA constancy rule. Unequivocal proof that the genes coding for 18S and 28S rRNA are amplified manyfold in the growing *Xenopus* oocyte has been provided by three different laboratories, all of them using molecular biological techniques (D. D. Brown and Dawid 1968, Evans and Birnstiel 1968, Gall 1968). The noted increase in rRNA gene number in *Xenopus* has since been found to take place in all amphibian species examined thus far, as well as in many insect oocytes and some other, phylogenetically distant species (see Tobler 1975, and Stark and Wahl 1984, for review). There was a long debate whether selective gene amplification does also occur in somatic cells for those genes which specify for a protein that is synthesized in large amounts in a given cell type. Spradling and Mahowald (1980) were the first to demonstrate that the genes coding for chorion proteins in *Drosophila* are amplified in ovarian follicle cells prior to their expression during oogenesis. In contrast to the amplification of rRNA-coding genes in oocytes, the amplified DNA sequences remain integrated in the chromosomes. The function of the amplification process, however, seems to be identical in both cases, namely to satisfy the demand for a large amount of a given gene product. Stress-induced gene amplification has been observed to occur in a variety of cell cultures when such cells are exposed to certain drugs or toxic compounds (see Stark and Wahl 1984, for a recent review). For example, cells cultured in the presence of methotrexate, an inhibitor of the enzyme dihydrofolate reductase (DHFR), gradually become resistant to the drug. Schimke et al. (1978) have shown that the increased resistance correlates with increased levels of the enzyme DHFR, which in turn is related to increased amounts of the gene coding for DHFR. Cultured cells increase their gene number gradually, possibly by gene duplications and selection for these cells. It is not known whether such stress-induced gene amplification plays any role in normal developmental processes.

A phenomenon related to gene amplification is gene magnification in *Drosophila* (Ritossa 1968), because both processes lead to an increase in the number of rRNA-coding genes. Gene magnification is defined as the process leading from subnormal doses of rDNA in bobbed mutants of *Drosophila* to normal amounts of rDNA by an increase in the content of rDNA over a few generations. The phenomenon takes place if bobbed mutants on the X chromosome with subnormal amounts of rDNA are placed opposite to a Y chromosome, partially or completely deficient in rDNA (Ritossa 1968). An increase in rRNA gene number has also been noted in *Drosophila* flies with genotypes containing only one nucleolus organizer per diploid set (Tartof 1971, Kunz and Schäfer 1976). This gene increase has been termed rDNA compensation by Tartof (1975). Spear and Gall (1973) questioned the existence of rDNA compensation, because in XO males of *D. melanogaster* an increased rDNA amount could not be found in the diploid brain but only in polytenic salivary gland cells. They therefore concluded that independent rDNA replication in polytene cells accounted for the total increase in rDNA measured in the DNA of whole XO flies by Tartof (1971). Grimm and Kunz (1980), however, demonstrated that at least in *D. hydei*, an increase in rRNA gene number does also occur in diploid tissues of genotypes containing only one nucleolus organizer. In a recent paper, Grimm et al. (1984) reported that either in larval brain ganglion cells or in adult thoracic muscle cells or in both, the rDNA content per nucleolus organizer is increased threefold in *D. hydei*, if particular regions of X or Y chromosomal heterochromatin are deleted. They introduced the term rDNA overreplication to designate the process of increase in rRNA gene number caused by deletions of particular heterochromatic regions on the sex chromosomes (Grimm et al. 1984). The two processes rDNA compensation and rDNA overreplication are undoubtedly related, if not identical. They may furthermore be related to gene magnification; however, unlike gene magnification in *D. melanogaster* (Ritossa 1968), rDNA overreplication does not restore the wild type in mutants which are of bobbed phenotype. There is also some evidence that rDNA overreplication is not a mechanism for regulating the number of rRNA genes (Grimm et al. 1984).

It seems to me an interesting fact that in *Drosophila*, which is a well-analyzed genus both at the genetical as well as at the cell and molecular biological level, so many phenomena have been found that lead to cell type-specific differences in DNA contents, some of which have been detected only by molecular biological techniques. The processes of polytenization, gene amplification, rDNA magnification, underreplication, and rDNA compensation (=rDNA overreplication?), all represent sensu stricto violations of the DNA constancy rule. One might therefore expect that far more exceptions to this rule should be detected if other species were investigated as thoroughly as *Drosophila*.

10.3 Gene Rearrangements

Despite the fact that chromatin elimination has been known for almost a century, biologists usually thought that this process represents a strange exception to the rule that all cells of an organism contain an identical set of genetic information and that the genome is quite stable. Even the genetic demonstration that genes in eukaryotic organisms may change their chromosomal location (=movable genetic elements) and that these rearrangements can affect the expression of other genes (McClintock 1967), remained largely unnoticed. However, with the discovery of transposable elements in bacteria and the possibility of cloning specific eukaryotic DNA sequences, further movable elements have been discovered in eukaryotes, first in *Drosophila* (Livak et al. 1978, Finnegan et al. 1978) and yeast (Cameron et al. 1979). Meanwhile, a great deal of information has accumulated about their structural organization, their behavior, and their effect on the expression of adjacent genes (see Shapiro 1983, for review). The transposition of these movable DNA elements has not been correlated with specific developmental events in any organism analyzed. Rather, it seems that the effect of transposons on neighboring genes is a random event caused by a chance rearrangement of the transposed element next to a given gene sequence. However, there is one good example known for the precise genetic rearrangement of specific DNA sequences during development, namely the rejoining of the structural genes for immunoglobulins during development of the immune system in bone marrow-derived lymphocytes of mammals (see Tonegawa 1983). Whereas the structural genes for the immunoglobulin polypeptide chain are scattered along a chromosome in the genome of a germ line cell, these gene segments become assembled by recombination during lymphocyte differentiation. Somatic recombination and a high mutation frequency in the amino-terminal coding region of the structural gene bring about the high degree of antibody diversity as observed in the mammalian immune system (Tonegawa 1983). In this case, therefore, the rearrangement of gene segments during development has known and important biological significance.

10.4 DNA Modification

Scarano et al. (1967) presented a highly speculative hypothesis according to which cells become different during embryonic development by specific alterations of the DNA base sequence. According to their scheme, the DNA of the germ line cells remains unaltered, whereas the somatic cells arise from mutation-like changes in the DNA sequences brought about by DNA-modifying enzymes. This hypothesis, in its essentials stating that cell differentiation is based on a series of directed mutations, has found little support. However, the occurrence of DNA modification by methylation of cytosine residues is a well-established fact in a great number of eukaryotic systems (see D. N. Cooper 1983, Doerfler 1983, Jaenisch and Jähner 1984, for recent reviews). There are many reports which show that methylation of certain cytosine residues is correlated with gene inactivation; on the other hand, no such correlation has been found in several other instances (also summarized in D. N. Cooper 1983). Furthermore, it is not clear whether differences in DNA methylation are the cause or the consequence of transcriptional activity, and if the former is really true, whether demethylation is a necessary rather than just a sufficient condition for transcription to occur. With respect to the subject of this chapter, i.e., the comparison of chromosome and chromatin elimination with other phenomena that lead to genome alterations, it is worth mentioning that the methylation pattern of germ line versus somatic DNA has also been determined in a number of cases. Such experiments demonstrated that germ line DNA contains much less methylated cytosine than DNA isolated from various somatic tissues of the same organism (see, e.g., Vanyushin et al. 1970, Kaput and Sneider 1979, Sturm and Taylor 1981). To my knowledge, the methylation pattern of germ line and somatic DNA has so far not been established for any higher metazoan species undergoing chromosome or chromatin elimination (for ciliates see Ammermann and Steinbrück 1981, and Steinbrück this Vol.).

10.5 Chromosome Inactivation

In Sect. 5.2 we have already seen that in the Coccidae, chromosome inactivation by heterochromatization and chromosome elimination represent two related phenomena, and that most likely the former process preceded the latter during evolution. The two processes also seem to be related in marsupial and placental mammals. However, whereas the function of these events in the coccids resides most likely in sex determination, it is generally thought that these processes in mammals serve as a dosage-compensation mechanism (see, e.g., Gartler and

Riggs 1983). There are further similarities between chromosome inactivation and elimination. First, both processes occur early during ontogenetic development. Whereas chromosome or chromatin diminution usually takes place in early cleavage stages (see Sect. 5.2 and Table 1), chromosome inactivation in the best-studied mammal, i.e., the house mouse, occurs at the morula to early blastocyst transition (see Gartler and Riggs 1983). Second, both events proceed in a heterogeneous way; i.e., in most cases chromosome elimination occurs in more than one cleavage stage, and chromosome inactivation takes place from the morula to the blastocyst stage, depending on the fate of the respective cells. The third similarity relies on the fact that both processes lead to an irreversible condition in the presomatic cells and their descendants. While chromosome inactivation by heterochromatization results in an irreversible inactivation of one of the two X chromosomes in the somatic cells of a female embryo, chromsome and chromatin elimination lead to an irreversible loss of genetic material in all presumptive somatic cells. The fundamental difference between the two processes consists, however, in the following: chromosome inactivation may, in principle, be a reversible process, which is not the case for chromosome or chromatin elimination, because the expelled genetic material is definitely lost. That the female germ cells indeed undergo an inactivation-reactivation cycle during early embryonic development, is now a well-established fact for the mouse (see McLaren 1981, Martin 1982, for review) as well as for some other mammals (cf. Gartler and Riggs 1983).

11 Summary and Concluding Remarks

The first and introducing chapter of this book served to summarize the classic data on germ line-soma segregation and differentiation, and to discuss these earlier findings with respect to recent results which have been obtained using modern techniques in cell and molecular biology. Special emphasis has been placed on a discussion of the chromatin diminution and chromosome elimination process, because this event represents the classical case that leads to genome differences in a multicellular organism. In the first four sections of this article I have presented some important classical theories and concepts, as well as some significant experimental findings in developmental biology. The terms chromatin and chromosome elimination and diminution have been redefined, and the new term ribonucleoprotein shedding was proposed for a special phenomenon which has hitherto been called chromatin elimination. The occurrence of chromatin and chromosome elimination in the animal kingdom has been reviewed. Both phenomena are phylogenetically widespread (cf. Table 1), but not universal. They usually take place during early cleavage stages in the presomatic nuclei, but may in a few cases also occur in the primordial germ cell nuclei, during spermatogenesis, or during mitotic parthenogenesis. In some instances it has been demonstrated that cytoplasmic factors play an important role in the prevention of chromosome and/or chromatin elimination. The functional aspects of chromatin and chromosome elimination, as well as the possible functions of the germ line-specific chromatin, have also been discussed. Molecular aspects of chromatin elimination in the nematode *Ascaris* are described in Sections 7 to 9. The amount of eliminated DNA has been determined and the eliminated DNA sequences characterized by density gradient analysis, renaturation kinetics, molecular hybridization experiments, and DNA cloning analysis. It has been shown that the eliminated chromatin in *Ascaris* contains a large amount of satellite DNA, but also nonsatellite DNA sequences. In the last section of this chapter, I have compared chromosome and chromatin elimination to other processes that lead to genome alterations, and presented evidence that chromosome inactivation and elimination are related phenomena.

The significance and mechanism of the chromosome and chromatin elimination process, as well as the genetic informational content of the eliminated DNA sequences, are not yet known. However, for those who have the privilege of being involved in elucidating these secrets, the phenomenon of chromosome and chromatin elimination is still as attractive and fascinating, as it already must have been about 100 years ago to the outstanding cell biologist Theodor Boveri.

Acknowledgments. I am particularly indebted to my present and past team for many helpful discussions, for their enthusiasm for doing research, for allowing me to cite unpublished work, and for creating a stimulating and friendly atmosphere. I am also very grateful to my colleagues and collaborators Drs. D. Meyer, F. Müller, J. Schowing, and R. Stocker for critical comments on the manuscript and for their valuable suggestions. Furthermore, I wish to thank the following colleagues, who have kindly accepted to read the manuscript, for their criticism, improvements, and suggestions: Drs. D. Ammermann, P. Borst, S. W. Emmons, J. G. Gall, P. Goldstein, W. Hennig, A. Lima-de-Faria, A. J. MacInnis, A. McLaren, K. B. Moritz, R. Nöthiger, S. Pimpinelli, E. Schierenberg, V. Schmid, J. Sommerville and S. Ward. The following colleagues kindly sent me reprints and preprints of unpublished work: Drs. D. Ammermann, R. Cassada, H. S. Chandra, S. W. Emmons, J. G. Gall, C. Goday, P. Goldstein, D. Hirsh, M. R. Klass, A. J. MacInnis, A. P. Mahowald, K. B. Moritz, T. Mutafova, E. Schierenberg, J. Sommerville, H. P. Stanley, J. E. Sulston, S. Ward, W. B. Wood. Many thanks go also to the Swiss National Science Foundation, since our research, partially described in this article, could not have been carried out without their financial support. I thank Mr. H. Gachoud for drawing the illustrations and for carrying out all the photographic work, and last but not least, I thank my wife Maedi Tobler, not just for typing the manuscript, but also for her patience during the elaboration of this review.

References

Aeby P (1979) Determination du nombre de gènes codant pour l'ARN de transfert dans les lignées germinale et somatique de l'*Ascaris lumbricoides* (var. *suum*). Dipl thesis, Univ Freiburg, Switzerland

Akifjew AP (1974) Silent DNA and its role in evolution (Russ.) Priroda 9:49–54

Albertson DG, Nwaorgu OC, Sulston JE (1979) Chromatin diminution and a chromosomal mechanism of sexual differentiation in *Strongyloides papillosus*. Chromosoma 75:75–87

Ammermann D (1965) Cytologische und genetische Untersuchungen an dem Ciliaten *Stylonychia mytilus* Ehrenberg. Arch Protistenkd 108:109–152

Ammermann D (1967) Die Cytologie der Parthenogenese bei dem Tardigraden *Hypsibius dujardini*. Chromosoma 23:203–213

Ammermann D (1971) Morphology and development of the macronuclei of the ciliates *Stylonychia mytilus* and *Euplotes aediculatus*. Chromosoma 33:209–238

Ammermann D (1986) Chromatin diminution and chromosome elimination: Mechanisms and adaptive significance. In: Cavalier-Smith T (ed) DNA and evolution: Natural selection and genome size. Wiley, New York, in press

Ammermann D, Steinbrück G (1981) Methylated bases in the DNA of the ciliate *Stylonychia mytilus*. Eur J Cell Biol 24:154–156

Andersen HA (1972) Induced elimination of DNA from macronucleus of *Tetrahymena pyriformis*. Exp Cell Res 74:610–613

Anya AO (1976) Physiological aspects of reproduction in nematodes. In: Dawes B (ed) Advances in parasitology, vol 14. Academic Press, London New York, pp 267–351

Back E, Müller F, Tobler H (1984a) Structural organization of the two main rDNA size classes of *Ascaris lumbricoides*. Nucleic Acids Res 12:1313–1332

Back E, Felder H, Müller F, Tobler H (1984b) Chromosomal arrangement of the two main rDNA size classes of *Ascaris lumbricoides*. Nucleic Acids Res 12:1333–1347

Back E, Van Meir E, Müller F, Schaller D, Neuhaus H, Aeby P, Tobler H (1984c) Intervening sequences in the ribosomal RNA genes of *Ascaris lumbricoides*: DNA sequences at junctions and genomic organization. EMBO J 3:2523–2529

Baltzer F (1962) Theodor Boveri. Wiss Verlagsges MBH, Stuttgart, pp 1–194

Banerjee MR (1959) Chromosome elimination during meiosis in the males of *Macroceroea* (Lohita) *grandis* (Gray) (Pyrrhocoridae, Heteroptera). Proc Zool Soc (Calcutta) 12:1–8

Bantock CR (1961) Chromosome elimination in cecidomyidae. Nature (London) 190:466–467

Bantock CR (1970) Experiments on chromosome elimination in the gall midge, *Mayetiola destructor*. J Embryol Exp Morphol 24:257–286

Basile R (1966) Estudo da espermatogênese e da ovogênes em *Rhynchosciara angelae* e da síntese de ácidos nucléicos e de proteínas no ovário. PhD thesis, Univ Sao Paulo, Sao Paulo

Bauer H (1932) Die Feulgensche Nuklealfärbung in ihrer Anwendung auf cytologische Untersuchungen. Z Zellforsch 15:225–247

Bauer H (1933) Die wachsenden Oocytenkerne einiger Insekten in ihrem Verhalten zur Nuklealfärbung. Z Zellforsch 18:254–298

Bauer H, Beermann W (1952a) Chromosomale Soma-Keimbahn-Differenzierung bei Chironomiden. Naturwissenschaften 39:22–23

Bauer H, Beermann W (1952b) Der Chromosomencyclus der Orthocladiinen (Nematocerca, Diptera). Z Naturforsch 7b:557–563

Beams HW, Kessel RG (1974) The problem of germ cell determinants. In: Bourne GH, Danielli JF, Jeon KW (eds) Int Rev Cytol Academic Press, London New York, pp 418–479

Beauchamp RS, Pasternak J, Straus NA (1979) Characterization of the genome of the free-living nematode *Panagrellus silusiae*: Absence of short period interspersion. Biochemistry 18:245–251

Beermann S (1959) Chromatin-Diminution bei Copepoden. Chromosoma 10:504–514

Beermann S (1977) The diminution of heterochromatic chromosomal segments in *Cyclops* (Crustacea, Copepoda). Chromosoma 60:297–344

Beermann S (1984) Circular and linear structures in chromatin diminution of *Cyclops*. Chromosoma 89:321–328

Beermann S, Meyer GF (1980) Chromatin rings as products of chromatin diminution in *Cyclops*. Chromosoma 77:277–283

Beermann W (1952) Chromomerenkonstanz und spezifische Modifikationen der Chromosomenstruktur in der Entwicklung und Organdifferenzierung von *Chironomus tentans*. Chromosoma 5:139–198

Beermann W (1972) Chromomeres and genes. In: Beermann W (ed) Developmental studies on giant chromosomes. Results and problems in cell differentiation, vol IV. Springer, Berlin Heidelberg New York, pp 1–33

Beneden E van (1983) Recherches sur la maturation de l'oeuf et la fécondation. Arch Biol 4:265–641

Bennett FD, Brown SW (1958) Life history and sex determination in the diaspine scale, *Pseudaulacaspis pentagona* (Targ.) (Coccoidea). Can Entomol 90:317–324

Bielka H, Schultz I, Böttger M (1968) Isolation and properties of DNA from eggs and gastrulae of *Ascaris lumbricoides*. Biochim Biophys Acta 157:209–212

Biocca E, Nascetti G, Iori A, Costantini R, Bullini L (1978) Descrizione di *Parascaris univalens*, parassita degli equini, e suo differenziamento da *Parascaris equorum*. Atti Accad Naz Lincei Cl Sci Fis Mat Nat Rend 65:133–141

Blackler AW (1958) Contribution to the study of germ-cells in the anura. J Embryol Exp Morphol 6:491–503

Bonnevie K (1902) Über Chromatindiminution bei Nematoden. Jena Z Naturwiss 36:275–288

Bostock C (1980) A function for satellie DNA? TIBS 5:117–119

Bounoure L (1939) L'origine des cellules reproductrices et le problème de la lignée germinale. Gauthier-Villars, Paris

Boveri T (1887) Über Differenzierung der Zellkerne während der Furchung des Eies von *Ascaris megalocephala*. Anat Anz 2:688–693

Boveri T (1888) Zellenstudien II. Die Befruchtung und Teilung des Eies von *Ascaris megalocephala*. Jena Z Naturwiss 22:685–882

Boveri T (1892) Über die Entstehung des Gegensatzes zwischen den Geschlechtszellen und den somatischen Zellen bei *Ascaris megalocephala*, nebst Bemerkungen zur Entwicklungsgeschichte der Nematoden. Sitzungsber Ges Morphol Physiol München 8:114–125

Boveri T (1899) Die Entwicklung von *Ascaris megalocephala* mit besonderer Rücksicht auf die Kernverhältnisse. In: Festschrift für C von Kupffer. Fischer, Jena, pp 383–430

Boveri T (1904) Ergebnisse über die Konstitution der chromatischen Substanz des Zellkerns. Jena, Fischer, pp 1–130

Boveri T (1910a) Über die Teilung centrifugierter Eier von *Ascaris megalocephala*. Arch Entwicklungsmech Org 30:101–125

Boveri T (1910b) Die Potenzen der Ascaris-Blastomeren bei abgeänderter Furchung. Zugleich ein Beitrag zur Frage qualitativ-ungleicher Chromosomen-Teilung. Festschr für R. Hertwig, vol III. Fischer, Jena, pp 131–214

Britten RJ, Kohne DE (1966) Nucleotide sequence repetition in DNA. Carnegie Inst Washington Yearb 65:78–106

Britten RJ, Kohne DE (1968) Repeated sequences in DNA. Science 161:529–540

Brown DD (1984) The role of stable complexes that repress and activate eucaryotic genes. Cell 37:359–365

Brown DD, Dawid IB (1968) Specific gene amplification in oocytes. Science 160:272–280

Brown SW, Bennett FD (1957) On sex determination in the diaspine scale *Pseudaulacaspis pentagona* (Targ.) (Coccoidea). Genetics 42:510–523

Brown SW, Chandra HS (1977) Chromosome imprinting and the differential regulation of homologous chromosomes. In: Goldstein L, Prescott DM (eds) Cell biology: A comprehensive treatise, vol I. Genetic mechanisms of cells. Academic Press, London New York, pp 109–189

Bueding E, Orrell SA (1964) A mild procedure for the isolation of polydisperse glycogen from animal tissues. J Biol Chem 239:4018–4020

Bullini L, Nascetti G, Ciafrè S, Rumore F, Biocca E (1978) Ricerche cariologiche ed elettroforetiche su *Parascaris univalens* e *Parascaris equorum*. Atti Accad Naz Lincei Cl Sci Fis Mat Nat Rend 65:151–159

Camenzind R (1971) The cytology of paedogenesis in the gall midge *Mycophila speyeri*. Chromosoma 35:393–402

Cameron JR, Loh EY, Davis RW (1979) Evidence for transposition of dispersed repetitive DNA families in yeast. Cell 16:739–751

Carter CE, Wells JR, MacInnis AJ (1972) DNA from anaerobic adult *Ascaris lumbricoides* and *Hymenolepis diminuta* mitochondria isolated by zonal centrifugation. Biochim Biophys Acta 262:135–144

Chitwood BG, Chitwood MB (1950) An introduction to nematology. Baltimore Monumental Printing, Baltimore, pp 1–213

Close RL (1984) Rates of sex chromosome loss during development in different tissues of the bandicoots *Perameles nasuta* and *Isoodon macrourus* (Marsupialia: Peramelidae). Aust J Biol Sci 37:53–61

Cooper DN (1983) Eukaryotic DNA methylation. Hum Genet 64:315–333

Cooper KW (1939) The nuclear cytology of the grass mite, *Pediculopsis graminum* (Reut.), with special reference to karyomerokinesis. Chromosoma 1:51–103

Croce CM, Kieba I, Koprowski H (1973) Unidirectional loss of human chromosomes in rat-human hybrids. Exp Cell Res 79:461–463

Crouse HV, Brown A, Mumford BC (1971) L-chromosome inheritance and the problem of chromosome "imprinting" in *Sciara* (Sciaridae, Diptera). Chromosoma 34:324–339

Dale PJ (1983) Protoplast culture and plant regeneration of cereals and other recalcitrant crops. Experientia Suppl 46:31–41

Darlington CD (1956) Chromosome Botany. Allen and Unwin, London, pp 1–186

Davey MR (1983) Recent developments in the culture and regeneration of plant protoplasts. Experientia Suppl 46:19–29

Davidson EH (1976) Gene activity in early development, 2nd edn. Academic Press, London New York, pp 1–452

Davidson EH, Hough BR, Amenson CS, Britten RJ (1973) General interspersion of repetitive with non-repetitive sequence elements in the DNA of Xenopus. J Mol Biol 77:1–23

Davis AH, Kidd GH, Carter CE (1979) Chromosome diminution in Ascaris suum. Two-fold increase of nucleosomal histone to DNA ratios during development. Biochim Biophys Acta 565:315–325

Dawson D, Buckley B, Cartinhour S, Myers R, Herrick G (1984) Elimination of germ-line tandemly repeated sequences from the somatic genome of the ciliate Oxytricha fallax. Chromosoma 90:289–294

DiBerardino MA, Hoffner YN, Etkin LD (1984) Activation of dormant genes in specialized cells. Science 224:946–952

Dixon KE (1981) The origin of the primordial germ cells in the amphibia. Neth J Zool 31:5–37

Doerfler W (1983) DNA methylation and gene activity. Annu Rev Biochem 52:93–124

DuBois AM (1932) Elimination of chromosomes during cleavage in the eggs of Sciara (Diptera). Proc Natl Acad Sci USA 18:352–356

DuBois AM (1933) Chromosome behavior during cleavage in the eggs of Sciara coprophila (Diptera) in the relation to the problem of sex determination. Z Zellforsch 19:595–614

Eddy EM (1975) Germ plasm and the differentiation of the germ cell line. In: Bourne GH, Danielli JF, Jeon KW (eds) International review of cytology. Academic Press, London New York, pp 229–280

Eddy EM, Clark JM, Gong D, Fenderson BA (1981) Review article: Origin and migration of primordial germ cells in mammals. Gamete Res 4:333–362

Edwards CL (1910a) The sex-determining chromosomes in Ascaris. Science 31:514–515

Edwards CL (1910b) The idiochromosomes in Ascaris megalocephala and Ascaris lumbricoides. Arch Zellforsch 5:422–429

Einsle U (1964) Die Gattung Cyclops s. str. im Bodensee. Arch Hydrobiol 60:133–199

Emmons SW, Yesner L (1984) High-frequency excision of transposable element Tcl in the nematode Caenorhabditis elegans is limited to somatic cells. Cell 36:599–605

Emmons SW, Klass MR, Hirsh D (1979) Analysis of the constancy of DNA sequences during development and evolution of the nematode Caenorhabditis elegans. Proc Natl Acad Sci USA 76:1333–1337

Emmons SW, Rosenzweig B, Hirsh D (1980) Arrangement of repeated sequences in the DNA of the nematode Caenorhabditis elegans. J Mol Biol 144:481–500

Emmons SW, Yesner L, Ruan K, Katzenberg D (1983) Evidence for a transposon in Caenorhabditis elegans. Cell 32:55–65

Endow SA, Gall JG (1975) Differential replication of satellite DNA in polyploid tissues of Drosophila virilis. Chromosoma 50:175–192

Evans D, Birnstiel ML (1968) Localization of amplified ribosomal DNA in the oocyte of Xenopus laevis. Biochim Biophys Acta 166:274–276

Fedoroff NV (1984) Transposable genetic elements in maize. Sci Am 250, 6:64–74

Felder H (1983) Lokalisierung von hochrepetitiven DNA-Sequenzen und ribosomalen Genen auf Chromosomen von Ascaris lumbricoides var. suum mittels in situ-Hybridisierungsexperimenten. Dipl thesis, Univ Freiburg, Switzerland, pp 1–54

Finch RA (1983) Tissue-specific elimination of alternative whole parental genomes in one barley hybrid. Chromosoma 88:386–393

Finnegan DJ, Rubin GM, Young MW, Hogness DS (1978) Repeated gene families in Drosophila melanogaster. Cold Spring Harbor Symp Quant Biol 42:1053–1063

Fogg LC (1930) A study of chromatin diminution in Ascaris and Ephestia. J Morphol 50:413

Fux T (1974) Chromosome elimination in Heteropeza pygmaea. II. Ultrastructure of the spindle apparatus. Chromosoma 49:99–112

Gall JG (1968) Differential synthesis of the genes for ribosomal RNA during amphibian oögenesis. Proc Natl Acad Sci USA 60:553–560

Gall JG (1981) Chromosome structure and the C-value paradox. J Cell Biol 91:3s–14s

Gartler SM, Riggs AD (1983) Mammalian X-chromosome inactivation. Annu Rev Genet 17:155–190

Geigy R (1931) Action de l'ultra-violet sur le pôle germinal dans l'oeuf de Drosophila melanogaster (Castration et mutabilité). Rev Suisse Zool 38:187–288

Gerbi S (1986) Chromosome mechanics in Sciarids (this Vol)

Gerhart JC (1980) Mechanisms regulating pattern formation in the amphibian egg and early embryo. In: Goldberger RF (ed) Biological regulation and development, vol II. Plenum Press, New York, pp 133–316

Geyer-Duszyńska I (1959) Experimental research on chromosome elimination in Cecidomyidae (Diptera). J Exp Zool 141:391–447

Geyer-Duszyńska I (1966) Genetic factors in oögenesis and spermatogenesis in Cecidomyiidae. In: Darlington CD, Lewis KR (eds) Chromosomes today, vol I. Oliver and Boyd, Edinburgh, pp 174–178

Goday C, Pimpinelli S (1984) Chromosome organization and heterochromatin elimination in Parascaris. Science 224:411–413

Goday C, Ciofi-Luzzatto A, Pimpinelli S (1985) Centromere ultrastructure in germ-line chromosomes of Parascaris. Chromosoma 91:121–125

Goldschmidt RB, Lin TP (1947) Chromatin diminution in Ascaris. Science 105:619

Goldstein P (1977) Chromatin diminution in early embryogenesis of Ascaris lumbricoides L. var. suum. J Morphol 152:141–151

Goldstein P (1978) Ultrastructural analysis of sex determination in Ascaris lumbricoides var. suum. Chromosoma 66:59–69

Goldstein P, Moens PB (1976) Karyotype analysis of Ascaris lumbricoides var. suum. Male and female pachytene nuclei by 3-D reconstruction from electron microscopy of serial sections. Chromosoma 58:101–111

Goldstein P, Straus NA (1978) Molecular characterization of Ascaris suum DNA and of chromatin diminution. Exp Cell Res 116:462–466

Goodrich HB (1916) The germ cells in Ascaris incurva. J Exp Zool 21:61–99

Goswami U (1973) Chromatin elimination in a rare species of nematode Physaloptera indiana. Curr Sci 42:576–577

Grimm C, Kunz W (1980) Disproportionate rDNA replication does occur in diploid tissue in Drosophila hydei. Mol Gen Genet 180:23–26

Grimm C, Kunz W, Franz G (1984) The organ-specific rRNA gene number in Drosophila hydei is controlled by sex heterochromatin. Chromosoma 89:48–54

Gurdon JB (1974) The control of gene expression in animal development. Clarendon Press, Oxford, pp 1–160

Gurdon JB, Uehlinger V (1966) "Fertile" intestine nuclei. Nature (London) 210:1240–1241

Gurdon JB, Laskey RA, Reeves OR (1975) The developmental capacity of nuclei transplanted from keratinized skin cells of adult frogs. J Embryol Exp Morphol 34:93–112

Hadorn E (1965) Problems of determination and transdetermination. In: Genetic control of differentiation. Brookhaven Natl Lab New York 18:148–159

Hadorn E (1978) Transdetermination. In: Ashburner M, Wright TRF (eds) The genetics and biology of Drosophila, vol II c. Academic Press, London New York, pp 555–617

Hatt P (1931) La fusion expérimentale d'oeufs de „Sabellaria alveolata L." et leur développement. Arch Biol 42:303–323

Hayman DL, Martin PG (1965) Sex chromosome mosaicism in the marsupial genera Isoodon and Perameles. Genetics 52:1201–1206

Hayman DL, Martin PG (1969) Cytogenetics of marsupials. In: Benirschke K (ed) Comparative mammalian cytogenetics. Springer, Berlin Heidelberg New York, pp 191–217

Hegner RW (1911) Germ-cell determinants and their significance. Am Nat 45:385–397

Hegner RW (1914) Studies on germ cells. J Morphol 25:375–509

Hennig W (1986) Heterochromatin and germ line limited DNA (this Vol)

Hennig W, Meer B (1971) Reduced polyteny of ribosomal RNA cistrons in giant chromosomes of Drosophila hydei. Nature (London) New Biol 233:70–72

Herla V (1893) Etude des variations de la mitose chez l'Ascaride mégalocéphale. Arch Biol 13:423–520

Hertwig R (1903) Über Korrelation von Zell- und Kerngröße und ihre Bedeutung für die geschlechtliche Differenzierung und die Teilung der Zelle. Biol Centralbl 23:49–62, 108–119

Hillman N, Sherman MI, Graham C (1972) The effect of spatial arrangement on cell determination during mouse development. J Embryol Exp Morphol 28:263–278

Hilscher W (1983) Problems of the Keimbahn. In: Hilscher W (ed) Problems of the Keimbahn, new work on mammalian germ cell lineage. Karger, Basel, pp 1–21

Hogue MJ (1910) Über die Wirkung der Centrifugalkraft auf die Eier von *Ascaris megalocephala*. Arch Entwicklungsmech Org 29:109–145

Huettner AF (1934) Octoploidy and diploidy in *Miastor americana*. Anat Rec 60 (Suppl):80

Illmensee K, Mahowald AP (1974) Transplantation of posterior polar plasm in *Drosophila*. Induction of germ cells at the anterior pole of the egg. Proc Natl Acad Sci USA 71:1016–1020

Illmensee K, Mahowald AP (1976) The autonomous function of germ plasm in a somatic region of the *Drosophila* egg. Exp Cell Res 97:127–140

Jaenisch R, Jähner D (1984) Methylation, expression and chromosomal position of genes in mammals. Biochim Biophys Acta 782:1–9

Jelinek WR, Schmid CW (1982) Repetitive sequences in eukaryotic DNA and their expression. Annu Rev Biochem 51:813–844

Ju-Chi-Li (1937) A six-chromosome *Ascaris* in Chinese horses. Science 86:101–102

Judd BH, Shen MW, Kaufman TC (1972) The anatomy and function of a segment of the X chromosome of *Drosophila melanogaster*. Genetics 71:139–156

Kahle W (1908) Die Paedogenesis der Cecidomyiden. Zoologica 21:1–80

Kaput J, Sneider TW (1979) Methylation of somatic vs germ cell DNAs analyzed by restriction endonuclease digestions. Nucleic Acids Res 7:2303–2322

Kaulenas MS, Fairbairn D (1968) RNA metabolism of fertilized *Ascaris lumbricoides* eggs during uterine development. Exp Cell Res 52:233–251

Kautzsch G (1913) Studien über Entwicklungsanomalien bei *Ascaris*. II. Arch Entwicklungsmech Org 35:642–691

Kilejian A, MacInnis AJ (1976) Density distribution of DNA from parasitic helminths with special reference to *Ascaris lumbricoides*. Rice Univ Stud 62:161–174

Kimble J, Hirsh D (1979) The postembryonic cell lineages of the hermaphrodite and male gonads in *Caenorhabditis elegans*. Dev Biol 70:396–417

King RL, Beams HW (1937) Effect of ultra-centrifuging on the egg of *Ascaris megalocephala*. Nature (London) 139:369–370

King RL, Beams HW (1938) An experimental study of chromatin diminution in *Ascaris*. J Exp Zool 77:425–443

Klingstedt H (1931) Digametie beim Weibchen der Trichoptere *Limnophilus decipiens* (Kol). Acta Zool Fenn 10:1–69

Kovaleva VG, Raikov IB (1978) Diminution and re-synthesis of DNA during development and senescence of the "diploid" macronuclei of the ciliate *Trachelonema sulcata* (Gymnostomata, Karyorelictida). Chromosoma 67:177–192

Krieg C, Cole T, Deppe U, Schierenberg E, Schmitt D, Yoder B, von Ehrenstein G (1978) The cellular anatomy of embryos of the nematode *Caenorhabditis elegans*. Dev Biol 65:193–215

Krüger E (1913) Fortpflanzung und Keimzellenbildung von *Rhabditis aberrans*, nov sp. Z Wiss Zool 105:87–124

Kühn A (1971) Lectures on developmental physiology, 2nd edn. Springer, Berlin Heidelberg New York, pp 1–535

Kuhn O, Tobler H (1978) Quantitative analysis of RNA, glycogen and nucleotides from different developmental stages of *Ascaris lumbricoides* (var. *suum*). Biochim Biophys Acta 521:251–266

Kunz W (1970) Genetische Aktivität der Keimbahnchromosomen während des Eiwachstums von Gallmücken (Cecidomyiidae). Verh Dtsch Zool Ges 1970:42–46

Kunz W, Schäfer U (1976) Variations in the number of the Y chromosomal rRNA genes in *Drosophila hydei*. Genetics 82:25–34

Kunz W, Trepte HH, Bier K (1970) On the function of the germ line chromosomes in the oogenesis of *Wachtliella persicariae* (Cecidomyiidae). Chromosoma 30:180–192

Landolt P, Tobler H (1980) The somatic DNA of *Ascaris lumbricoides* shows short-period interspersion. Experientia 36:750

Laskey RA, Gurdon JB (1970) Genetic content of adult somatic cells tested by nuclear transplantation from cultured cells. Nature (London) 228:1332–1334

Laufer JS, Ehrenstein G von (1981) Nematode development after removal of egg cytoplasm: Absence of localized unbound determinants. Science 211:402–405

Laufer JS, Bazzicalupo P, Wood WB (1980) Segregation of developmental potential in early embryos of *Caenorhabditis elegans*. Cell 19:569–577

Lauth MR, Spear BB, Heumann J, Prescott DM (1976) DNA of ciliated protozoa: DNA sequence diminution during macronuclear development of *Oxytricha*. Cell 7:67–74

Levine M, Garen A, Lepesant JA, Lepesant-Kejzlarova J (1981) Constancy of somatic DNA organization in developmentally regulated regions of the *Drosophila* genome. Proc Natl Acad Sci USA 78:2417–2421

Lewin R (1982) Repeated DNA still in search of a function. Science 217:621–623

Liao LW, Rosenzweig B, Hirsh D (1983) Analysis of a transposable element in *Caenorhabditis elegans*. Proc Natl Acad Sci USA 80:3585–3589

Lima-de-Faria A (1975) The relation between chromomeres, replicons, operons, transcription units, genes, viruses and palindromes. Hereditas 81:249–284

Lima-de-Faria A (1983) Molecular evolution and organization of the chromosome. Elsevier, Amsterdam, pp 1–1186

Lin TP (1954) The chromosomal cycle in *Parascaris equorum* (*Ascaris megalocephala*): Oogenesis and diminution. Chromosoma 6:175–198

Livak KJ, Freund R, Schweber M, Wensink PC, Meselson M (1978) Sequence organization and transcription at two heat shock loci in *Drosophila*. Proc Natl Acad Sci USA 75:5613–5617

Mahowald AP (1968) Polar granules of *Drosophila*. II. Ultrastructural changes during early embryogenesis. J Exp Zool 167:237–262

Mahowald AP (1971 a) Polar granules of *Drosophila*. IV. Cytochemical studies showing loss of RNA from polar granules during early stages of embryogenesis. J Exp Zool 176:345–352

Mahowald AP (1971 b) Origin and continuity of polar granules. In: Reinert J, Ursprung H (eds) Results and problems in cell differentiation, vol II. Origin and continuity of cell organelles. Springer, Berlin Heidelberg New York, pp 158–169

Mahowald AP, Allis CD, Karrer KM, Underwood EM, Waring GL (1979) Germ plasm and pole cells of *Drosophila*. In: Subtelny S, Konigsberg IR (eds) Determinants of spatial organization. Academic Press, London New York, pp 127–146

Mangold O (1923) Transplantationsversuche zur Frage der Spezifität und der Bildung der Keimblätter bei Triton. Arch Mikrosk Anat Entwicklungsmech 100:198–301

Mangold O, Seidel F (1927) Homoplastische und heteroplastische Verschmelzung ganzer Tritonkeime. Wilhelm Roux' Arch Entwicklungsmech Org 111:593–665

Markert CL, Petters RM (1978) Manufactured hexaparental mice show that adults are derived from three embryonic cells. Science 202:56–58

Martin GR (1982) X-chromosome inactivation in mammals. Cell 29:721–724

Matuszewski B (1982) Diptera I: Cecidomyiidae. In: John B (ed) Animal cytogenetics, vol III. Insecta 3. Borntraeger, Berlin, pp 1–140

McCarthy BJ, Hoyer BH (1964) Identity of DNA and diversity of messenger RNA molecules in normal mouse tissues. Proc Natl Acad Sci USA 52:915–922

McClintock B (1967) Genetic systems regulating gene expression during development. Dev Biol Suppl 1:84–112

McLaren A (1981) Germ cells and soma: a new look at an old problem. Yale Univ Press, New Haven, pp 1–119

McLaren A, Wylie CC (1983) Current problems in germ cell differentiation. Cambridge Univ Press, Cambridge, pp 1–401

McTavish C, Sommerville J (1980) Macronuclear DNA organization and transcription in *Paramecium primaurelia*. Chromosoma 78:147–164

Metz CW (1938) Chromosome behavior, inheritance and sex determination in *Sciara*. Am Nat 72:485–520

Metz CW, Lawrence EG (1938) Preliminary observations on *Sciara* hybrids. J Hered 29:179–186

Meyer GF, Lipps HJ (1980) Chromatin elimination in the hypotrichous ciliate *Stylonychia mytilus*. Chromosoma 77:285–297

Meyer O (1895) Celluläre Untersuchungen an Nematoden-Eiern. Jena Z Naturwiss 29:391–410

Mintz B (1962) Formation of genotypically mosaic mouse embryos. Am Zool 2:432

Moritz KB (1967a) Die Blastomerendifferenzierung für Soma and Keimbahn bei *Parascaris equorum*. I. Cytochemische und photometrische Untersuchungen. Wilhelm Roux' Arch Entwicklungsmech 159:31–88

Moritz KB (1967b) Die Blastomerendifferenzierung für Soma und Keimbahn bei *Parascaris equorum*. II. Untersuchungen mittels UV-Bestrahlung und Zentrifugierung. Wilhelm Roux' Arch Entwicklungsmech 159:203–266

Moritz KB (1970a) Quantitative aspects of chromosomal composition in *Ascaris megalocephala*. In: Wied GL, Bahr GF (eds) Introduction to quantitative cytochemistry, vol II. Academic Press, London New York, pp 57–75

Moritz KB (1970b) DNS-Variation im keimbahnbegrenzten Chromatin und autoradiographische Befunde zu seiner Funktion bei *Parascaris equorum*. Verh Dtsch Zool Ges 64:36–42

Moritz KB (1977) Die Chromosomen von *Ascaris* in der Keimbahn und im embryonalen Soma. Verh Dtsch Zool Ges 1977:290

Moritz KB (1984) Der molekulare Mechanismus der Chromatindiminution. Zugleich ein Beitrag zur Telomerorganisation und -dynamik. Verh Dtsch Zool Ges 1984 77:197

Moritz KB, Bauer M (1984) Cytoplasmatische Bedingungen der Chromatindiminution bei *Parascaris equorum*. Verh Dtsch Zool Ges 77:164

Moritz KB, Roth GE (1976) Complexity of germline and somatic DNA in *Ascaris*. Nature (London) 259:55–57

Moritz KB, Roth GE (1978) Die mitochondriale DNA im Ei von *Ascaris*. Verh Dtsch Zool Ges 71:231

Müller F, Walker P, Aeby P, Neuhaus H, Back E, Tobler H (1982a) Molecular cloning and sequence analysis of highly repetitive DNA sequences contained in the eliminated genome of *Ascaris lumbricoides*. In: Burger MM, Weber R (eds) Embryonic development, part A: Genetic aspects. Liss, New York, pp 127–138

Müller F, Walker P, Aeby P, Neuhaus H, Felder H, Back E, Tobler H (1982b) Nucleotide sequence of satellite DNA contained in the eliminated genome of *Ascaris lumbricoides*. Nucleic Acids Res 10:7493–7510

Müntzing A (1949) Accessory chromosomes in *Secale* and *Poa*. Hereditas Suppl 35:402–411

Murti KG, Prescott DM (1970) Micronuclear ribonucleic acid in *Tetrahymena pyriformis*. J Cell Biol 47:460–467

Mutafova T (1975) Morphology and behaviour of sex chromosomes during meiosis in *Ascaris suum*. Z Parasitenkd 46:291–295

Nagl W (1976) Zellkern und Zellzyklen. Ulmer, Stuttgart, pp 1–486

Nagl W (1978) Endopolyploidy and polyteny in differentiation and evolution. Elsevier/North-Holland, Amsterdam, pp 1–283

Nagl W (1983) Heterochromatin elimination in the orchid *Dendrobium*. Protoplasma 118:234–237

Nelson-Rees WA, Hoy MA, Roush RT (1980) Heterochromatinization, chromatin elimination and haploidization in the parahaploid mite *Metaseiulus occidentalis* (Nesbitt) (Acarina: Phytoseiidae). Chromosoma 77:263–276

Nicklas RB (1959) An experimental and descriptive study of chromosome elimination in *Miastor* spec. (Cecidomyidae; Diptera). Chromosoma 10:301–336

Nicklas RB (1960) The chromosome cycle of a primitive Cecidomyiid *Mycophila* speyeri. Chromosoma 11:402–418

Nieuwkoop PD, Sutasurya LA (1979) Primordial germ cells in the chordates. Cambridge Univ Press, Cambridge, pp 1–187

Nieuwkoop PD, Sutasurya LA (1981) Primordial germ cells in the invertebrates. Cambridge Univ Press, Cambridge, pp 1–258

Nigon V (1965) Développement et reproduction des Nématodes. In: Grassé PP (ed) Traité de zoologie, vol IV. Masson, Paris, pp 218–386

Nigon V, Guerrier P, Monin H (1960) L'architecture polaire de l'oeuf et les mouvements des constituants cellulaires au cours des premières étapes du développement chez quelques Nématodes. Bull Biol Fr Belg 94:131–202

Noda K, Kanai C (1977) An ultrastructural observation on *Pelmatohydra robusta* at sexual and asexual stages, with a special reference to "germinal plasm". J Ultrastruct Res 61:284–294

Noll F, Bielka H (1968) Zur Biochemie der Embryogenese von *Ascaris* I. Gehalt und Verteilungsmuster der Nukleinsäuren. Acta Biol Med Germ 20:565–575

Nur U (1980) Evolution of unusual chromosome systems in scale insects (Coccoidea: Homoptera). In: Blackman RL, Hewitt GM, Ashburner M (eds) Insect cytogenetics. Blackwell Sci Publ, Oxford, pp 97–117

Nussbaum M (1880) Zur Differenzierung des Geschlechts im Thierreich. Arch Mikrosk Anat Entwicklungsmech 18:1–120

Okada TS (1983) Recent progress in studies of the transdifferentiation of eye tissue in vitro. Cell Differ 13:177–183

Painter TS (1945) Chromatin diminution. Trans Conn Acad Arts Sci 36:443–448

Painter TS (1966) The role of the E-chromosomes in *Cecidomyiidae*. Proc Natl Acad Sci USA 56:853–855

Painter TS, Stone W (1935) Chromosome fusion and speciation in *Drosophilae*. Genetics 20:327–341

Pasteels J (1948a) Etude cytochimique des acides nucléiques dans le cycle germinal de l'*Ascaris megalocephala*. Experientia 4:150–152

Pasteels J (1948b) Recherches sur le cycle germinal chez l'*Ascaris*. Etude cytochimique des acides nucléiques dans l'oögénèse, la spermatogénèse et le développement chez *Parascaris equorum* Goerze. Arch Biol 59:405–446

Pasternak J, Barrell R (1976) Quantitation of nuclear DNA in *Ascaris lumbricoides*: DNA constancy and chromatin diminution. Genet Res Cambridge 27:339–348

Radzikowski S (1973) Die Entwicklung des Kernapparates und die Nukleinsäuresynthese während der Konjugation von *Chilodonella cucullulus* O. F. Müller. Arch Protistenkd 115:419–428

Reitberger A (1934) Das Verhalten der Chromosomen bei der pädogenetischen Entwicklung der Cecidomyide *Oligarces paradoxus*, mit besonderer Berücksichtigung der Chromosomen-Elimination. Verh Schweiz Naturforsch Ges 115:359–360

Reitberger A (1940) Die Cytologie des pädagenetischen Entwicklungszyklus der Gallmücke *Oligarces paradoxus* Mein. Chromosoma 1:391–473

Rempel JG, Church NS (1969) The embryology of *Lytta viridana* Le Conte (Coleoptera: Meloidae) IV. Chromatin elimination. Can J Zool 47:351–353

Rieger R, Michaelis A, Green MM (1968) A glossary of genetics and cytogenetics. Springer, Berlin Heidelberg New York, pp 1–507

Ris H, Kleinfeld R (1952) Cytochemical studies on the chromatin elimination in *Solenobia* (Lepidoptera). Chromosoma 5:363–371

Ritossa FM (1968) Unstable redundancy of genes for ribosomal RNA. Proc Natl Acad Sci USA 60:509–516

Rosenzweig B, Liao LW, Hirsh D (1983a) Sequence of the *C. elegans* transposable element Tc1. Nucleic Acids Res 11:4201–4209

Rosenzweig B, Liao LW, Hirsh D (1983b) Target sequences for the *C. elegans* transposable element Tc1. Nucleic Acids Res 11:7137–7140

Roth GE (1979) Satellite DNA properties of the germ line limited DNA and the organization of the somatic genomes in the nematodes *Ascaris suum* and *Parascaris equorum*. Chromosoma 74:355–371

Roth GE, Moritz KB (1981) Restriction enzyme analysis of the germ line limited DNA of *Ascaris suum*. Chromosoma 83:169–190

Ruan K, Emmons SW (1984) Extrachromosomal copies of transposon Tc1 in the nematode *Caenorhabditis elegans*. Proc Natl Acad Sci USA 81:4018–4022

Sang JH (1984) Genetics and development. Longman, London, pp 1–398

Scarano E, Iaccarino M, Grippo P, Parisi F (1967) The heterogeneity of thymine methyl group origin in DNA pyrimidine isostichs of developing sea urchin embryos. Proc Natl Acad Sci USA 57:1394–1400

Schachat F, O'Connor DJ, Epstein HF (1978) The moderately repetitive DNA sequences of *Caenorhabditis elegans* do not show short-period interspersion. Biochim Biophys Acta 520:688–692

Schimke RT, Alt FW, Kellems RE, Kaufman RJ, Bertino JR (1978) Amplification of folate reductase genes in methotrexate-resistant cultured mouse cells. Cold Spring Harbor Symp Quant Biol 42:649–657

Schleip W (1923) Die Wirkung des ultravioletten Lichtes auf die morphologischen Bestandteile des *Ascaris*eies. Arch Zellforsch 17:289–367

Schmid V, Alder H (1984) Isolated mononucleated, striated muscle can undergo pluripotent transdifferentiation and form a complex regenerate. Cell 38:801–809

Schrader F (1935) Notes on the mitotic behavior of long chromosomes. Cytologia 6:422–430

Schwartz V, Meister H (1975) Die Extinktion der feulgengefärbten Makronucleusanlage von *Paramaecium bursaria* in der DNS-armen Phase. Arch Protistenkd 117:60–64

Searcy DG, MacInnis AJ (1970) Measurements by DNA renaturation of the genetic basis of parasitic reduction. Evolution 24:796–806

Seiler J (1914) Das Verhalten der Geschlechtschromosomen bei Lepidopteren. Arch Zellforsch 13:159–269

Shapiro JA (1983) Mobile genetic elements. Academic Press, London New York, pp 1–688

Singer MF (1982) Highly repeated sequenes in mammalian genomes. Int Rev Cytol 76:67–112

Slack JMW (1983) From egg to embryo: Determinative events in early development. Cambridge Univ Press, Cambridge, pp 1–241

Smith LD (1966) The role of a "germinal plasm" in the formation of primordial germ cells in *Rana pipiens*. Dev Biol 14:330–347

Sonneborn TM (1974) *Paramecium aurelia*. In: King RC (ed) Handbook of genetics, vol II. Plenum Press, New York, pp 469–594

Sonnenblick BP (1965) The early embryology of *Drosophila melanogaster*. In: Demerec M (ed) Biology of *Drosophila*. Hafner, New York, pp 62–167

Sorsa M, Suomalainen E (1975) Electron microscopy of chromatin elimination in *Cidaria* (Lepidoptera). Hereditas 80:35–40

Spear BB, Gall JG (1973) Independent control of ribosomal gene replication in polytene chromosomes of *Drosophila melanogaster*. Proc Natl Acad Sci USA 70:1359–1363

Spemann H (1903) Entwicklungsphysiologische Studien am *Triton*-Ei. II. Arch Entwicklungsmech Org 15:448–534

Spradling AC, Mahowald AP (1980) Amplification of genes for chorion proteins during oogenesis in *Drosophila melanogaster*. Proc Natl Acad Sci USA 77:1096–1100

Stanley HP, Kasinsky HE, Bols NC (1984) Meiotic chromatin diminution in a vertebrate, the holocephalan fish *Hydrolagus colliei* (Chondrichthyes, Holocephali). Tissue Cell 16:203–215

Stark GR, Wahl GM (1984) Gene amplification. Annu Rev Biochem 53:447–491

Steinbrück G (1986) Molecular reorganization during nuclear differentiation in ciliates (this Vol)

Steinbrück G, Haas I, Hellmer KH, Ammermann D (1981) Characterization of macronuclear DNA in five species of Ciliates. Chromosoma 83:199–208

Stevens NM (1909) The effect of ultraviolet-light upon the developing eggs of *Ascaris megalocephala*. Arch Entwicklungsmech Org 27:622–639

Steward FC, Mapes MO, Kent AE, Holsten RD (1964) Growth and development of cultured plant cells. Science 143:20–27

Strassen O zur (1896) Embryonalentwicklung der *Ascaris megalocephala*. Arch Entwicklungsmech 3:27–105, 133–190

Strassen O zur (1898) Über die Riesenbildung bei *Ascaris*eiern. Arch Entwicklungsmech 7:642–676

Strassen O zur (1906) Die Geschichte der T-Riesen von *Ascaris megalocephala* als Grundlage zu einer Entwicklungsmechanik dieser Spezies. Zoologica 40:1–342

Streble H, Krauter D (1982) Das Leben im Wassertropfen, Mikroflora und Mikrofauna des Süßwassers. Franckh'sche Verlagshandlung, Keller u Cie, Stuttgart, pp 1–336

Streeck RE, Moritz KB, Beer K (1982) Chromatin diminution in *Ascaris suum*: nucleotide sequence of the eliminated satellite DNA. Nucleic Acids Res 10:3495–3502

Strome S, Wood WB (1982) Immunofluorescence visualization of germ-line-specific cytoplasmic granules in embryos, larvae, and adults of *Caenorhabditis elegans*. Proc Natl Acad Sci USA 79:1558–1562

Strome S, Wood WB (1983) Generation of asymmetry and segregation of germ-line granules in early *C. elegans* embryos. Cell 35:15–25

Sturm KS, Taylor JH (1981) Distribution of 5-methylcytosine in the DNA of somatic and germ-line cells from bovine tissues. Nucleic Acids Res 9:4537–4546

Sulston JE, Brenner S (1974) The DNA of *Caenorhabditis elegans*. Genetics 77:95–104

Sulston JE, Schierenberg E, White JG, Thomson JN (1983) The embryonic cell lineage of the nematode *Caenorhabditis elegans*. Dev Biol 100:64–119

Tadano M (1968) Nemathelminthes. In: Kumé M, Dan K (eds) Invertebrate embryology. Nolit, Belgrad, pp 159–191

Tanabe K, Kotani M (1974) Relationship between the amount of the "germinal plasm" and the number of primordial germ cells in *Xenopus laevis*. J Embryol Exp Morphol 31:89–98

Tardent P (1978) Coelenterata, Cnidaria. In: Seidel F (ed) Morphogenese der Tiere. Fischer, Jena, pp 69–415

Tarkowski AK (1961) Mouse chimaeras developed from fused eggs. Nature (London) 190:857–860

Tartof KD (1971) Increasing the multiplicity of ribosomal RNA genes in *Drosophila melanogaster*. Science 171:294–297

Tartof KD (1975) Redundant genes. Annu Rev Genet 9:355–385

Tchou Su, Chen-Chao-Hsi (1937) Une nouvelle race chinoise d'*Ascaris megalocephala* (type trivalens). C R Acad Sci :1676–1677

Thomas C, Prasad RS (1980) Chromosome elimination in *Ctenocephalides orientis* (Siphonaptera). Cytobios 29:109–114

Thomas HM, Pickering RA (1983) Chromosome elimination in *Hordeum vulgare* × *H. bulbosum* hybrids. 1. Comparison of stable and unstable amphidiploids. Theor Appl Genet 66:135–140

Tobler H (1972) The problem of genetic identity of different cell types. In: Ursprung H (ed) Nucleic acid hybridization in the study of cell differentiation. Results and problems in cell differentiation, vol III. Springer, Berlin Heidelberg New York, pp 1–9

Tobler H (1975) Occurrence and developmental significance of gene amplification. In: Weber R (ed) The biochemistry of animal development, vol III. Academic Press, London New York, pp 91–143

Tobler H (1976) Genetic difference between germ line and somatic DNA in *Ascaris lumbricoides*. In: Müller-Bérat N, Rosenfeld C, Tarin D, Viza D (eds) Progress in differentiation research. Elsevier/North-Holland, Amsterdam, pp 147–154

Tobler H, Gut C (1974) Mitochondrial DNA from 4-cell stages of *Ascaris lumbricoides*. J Cell Sci 16:593–601

Tobler H, Smith KD, Ursprung H (1972) Molecular aspects of chromatin elimination in *Ascaris lumbricoides*. Dev Biol 27:190–203

Tobler H, Zulauf E, Kuhn O (1974) Ribosomal RNA genes in germ line and somatic cells of *Ascaris lumbricoides*. Dev Biol 41:218–223

Tobler H, Müller F, Back E, Aeby P (1985) Germ line – soma differentiation in *Ascaris*: A molecular approach. Experientia 41:1311–1319

Tonegawa S (1983) Somatic generation of antibody diversity. Nature (London) 302:575–581

Triantaphyllou AC (1971) Genetics and cytology. In: Zuckermann BM, Mai WF, Rohde RA (eds) Plant parasitic nematodes, vol II. Cytogenetics, host-parasite interactions and physiology. Academic Press, London New York, pp 1–34

Ubisch L von (1943) Über die Bedeutung der Diminution von *Ascaris megalocephala*. Acta Biotheor 7:163–181

Ueda R, Okada M (1982) Induction of pole cells in sterilized *Drosophila* embryos by injection of subcellular fractions from eggs. Proc Natl Acad Sci USA 79:6946–6950

Vanyushin BF, Tkacheva SG, Belozersky AN (1970) Rare bases in animal DNA. Nature (London) 225:948–949

Vassilev I, Mutafova T (1974) Comparative studies on the karyotype of *Ascaris suum* and "*Ascaris ovis*". Z Parasitenkd 43:115–121

Vendrely R (1955) The deoxyribonucleic acid content of the nucleus. In: Chargaff E, Davidson JN (eds) The nucleic acids, vol II. Academic Press, London New York, pp 155–180

Vereiskaya VN (1975) A cytochemical study of the elimination chromatin in the silkworm (*Bombyx mori* L.) meiosis. Tsitologyia 17:603–606

Wakahara M (1977) Partial characterization of "primordial germ cell-forming activity" localized in vegetal pole cytoplasm in anuran eggs. J Embryol Exp Morphol 39:221–233

Wakahara M (1978) Induction of supernumerary primordial germ cells by injecting vegetal pole cytoplasm into *Xenopus* eggs. J Exp Zool 203:159–164

Wallace H, Morray J, Langridge WHR (1971) Alternative model for gene amplification. Nature (London) New Biol 230:201–203

Walton AC (1917) The oogenesis and early embryology of *Ascaris canis* Werner. J Morphol 30:527–603

Walton AC (1924) Studies on nematode gametogenesis. Z Zellen- Gewebel 1:167–239

Walton AC (1959) Some parasites and their chromosomes. J Parasitol 45:1–20

Walton AC (1974) Gametogenesis. In: Chitwood BG, Chitwood MB (eds) Introduction to nematology. Univ Park Press, Baltimore, pp 191–201

Waring GL, Allis CD, Mahowald AP (1978) Isolation of polar granules and the identification of polar granule-specific protein. Dev Biol 66:197-206

Weismann A (1885) Die Continuität des Keimplasmas als Grundlage einer Theorie der Vererbung. Fischer, Jena

Weismann A (1892) Das Keimplasma. Eine Theorie der Vererbung. Fischer, Jena

Weiss MC, Green H (1967) Human-mouse hybrid cell lines containing partial complements of human chromosomes and functioning human genes. Proc Natl Acad Sci USA 58:1104–1111

White MJD (1936) Chromosome cycle of *Ascaris megalocephala*. Nature (London) 137:783

White MJD (1947) The cytology of the Cecidomyidae (Diptera) III. J Morphol 80:1–24

White MJD (1950) Cytological studies on gall midges (Cecidomyidae). Univ Tex Publ 5007:1–80

White MJD (1954) Animal cytology and evolution, 2nd edn. Cambridge Univ Press, Cambridge, pp 1–454

White MJD (1973) Animal cytology and evolution, 3rd edn. Cambridge Univ Press, Cambridge, pp 1–961

Whitington PMcD, Dixon KE (1975) Quantitative studies of germ plasm and germ cells during early embryogenesis of *Xenopus laevis*. J Embryol Exp Morphol 33:57–74

Wilson EB (1896) The cell in development and inheritance, 1st edn. Macmillan, New York, pp 1–371

Wilson EB (1925) The cell in development and heredity, 3rd edn. Macmillan, New York, pp 1–1232

Wolf N, Priess J, Hirsh D (1983) Segregation of germline granules in early embryos of *Caenorhabditis elegans*: an electron microscopic analysis. J Embryol Exp Morphol 73:297–306

Wolff E (1964) L'origine de la lignée germinale. Hermann, Paris, pp 1–370

Wood WB, Strome S, Laufer JS (1983) Localization and determination in embryos of *Caenorhabditis elegans*. In: Jeffery WR, Raff RA (eds) Time, space, and pattern in embryonic development. Liss, New York, pp 221–239

Yamada T, McDevitt DS (1984) Conversion of iris epithelial cells as a model of differentiation control. Differentiation 27:1–12

Yamaguchi Y, Murakami K, Furusawa M, Miwa J (1983) Germline-specific antigens identified by monoclonal antibodies in the nematode *Caenorhabditis elegans*. Dev Growth Differ 25:121–131

Yao MC, Gall JG (1979) Alteration of the *Tetrahymena* genome during nuclear differentiation. J Protozool 26:10–13

Yao MC, Gorovsky MA (1974) Comparison of the sequences of macro- and micronuclear DNA of *Tetrahymena pyriformis*. Chromosoma 48:1–18

Yao T, Pai S (1942) Heteropycnosis and chromatin diminution in *Cosmocerca* sp. Sci Rec Acad Sinica 1:197–202

Zalokar M (1976) Autoradiographic study of protein and RNA formation during early development of *Drosophila* eggs. Dev Biol 49:425–437

Unusual Chromosome Movements in Sciarid Flies

S. A. GERBI [1]

1 Introduction

1.1 Background

Sciarid flies are Diptera (two-winged flies) of the suborder Nematocera; the family Sciaridae are commonly known as fungus gnats, since they eat mushrooms and other fungi. Morphological characteristics of *Sciara* species useful for taxonomic purposes were first tabulated by Johannsen (1909, 1912). More recently, the family of *Sciara* has been renamed *Bradysia* (Steffan 1966), but since almost none of the researchers who work with this family have switched over to the new name, I will use the original name of *Sciara* in this review.

In the early 1920's Dr. Charles W. Metz obtained *Sciara* from the pigeon house at the Cold Spring Harbor laboratory in New York, and began laboratory stocks that ever since have been a fruitful source of material for cytogenetic research. *Sciara* has many features making it unique for such research. The beautiful polytene chromosomes of the salivary glands of *Sciara*, which have undergone a few more rounds of endoduplication than those of *Drosophila*, will not be discussed in this review; suffice it to say that the *Sciara* polytene chromosomes possess DNA puffs (not found in *Drosophila* or *Chironomus*) that occur in the salivary glands in late larval life and are sites of DNA amplification. The focus of this review will be on the unusual chromosome movements found in the embryonic soma and germ line of *Sciara*; some aspects of this topic have been reviewed previously by others (Metz 1938, White 1973). I will first give an overview of the main events of the unique chromosome cycle of *Sciara*, and then discuss each part of this cycle in more detail in subsequent sections. The majority of information derives from the species *Sciara coprophila*, which will be used as the example throughout this review except where otherwise indicated. However, the main features to be described for *Sciara coprophila* are also generally true for other Sciarid species, too.

1.2 The Chromosome Complement

1.2.1 The "Ordinary" Chromosomes

The chromosome complement found in *Sciara coprophila* is shown in Fig. 1. There are three pairs of autosomes: chromosome II is a short acrocentric rod,

---

[1] Brown University, Division of Biology and Medicine, Providence, Rhode Island 02912, USA.

Results and Problems in Cell Differentiation 13
Germ Line – Soma Differentiation (Ed. by W. Hennig)
© Springer-Verlag Berlin Heidelberg 1986

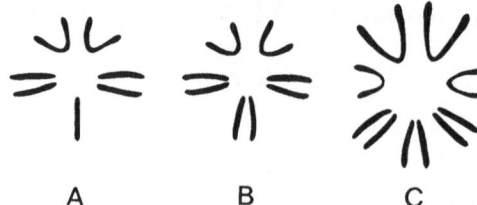

Fig. 1. Schematic drawings of the *S. coprophila* chromosome set in **A** male somatic cells, which are XO, **B** female somatic cells, which are XX, and **C** germ cells of both sexes, which are XX and contain L chromosomes. (Taken from DuBois 1933)

chromosome III is a somewhat longer acrocentric rod and chromosome IV is a metacentric V-shaped chromosome. Female somatic cells also contain two X chromosomes, which are acrocentric rods (Fig. 1 B), while male somatic cells have only a single X chromosome in addition to the autosomes (Fig. 1 A). Note from Fig. 1 A that there is no Y chromosome in *Sciara*; the male soma is XO in constitution. Table 1 summarizes the genetic markers used to establish these linkage groups; the only markers still extant are swollen, petite, and Wavy, which are all on the sex chromosomes.

Table 1. Linkage groups of *Sciara coprophila*

Chromosome	Mutation	Reference
II	truncate (rec.)	Metz (1926b)
	Dash (dom.)	Smith-Stocking (1936)
	Varied (dom.)	Smith-Stocking (1936)
	Stop (dom.)[a]	Crouse and Smith-Stocking (1938)
III	Curly (dom.)	Smith-Stocking (1936)
IV	Blister (dom.)	Smith-Stocking (1936)
	Delta (dom.)	Smith-Stocking (1936)
	Fused (dom.)	Smith-Stocking (1936)
	Stop (dom.)[a]	Crouse and Smith-Stocking (1938)
X	swollen (rec.)	Metz and Ullian (1929)
	narrow (rec.)	Metz and Schmuck (1929a, b; 1931a)
	miniature (rec.)	Smith-Stocking (1936)
	round (rec.)	Smith-Stocking (1936)
	petite (rec.)	Crouse (1961)
X'	Wavy (dom.)	Metz and Smith (1931)
L	No markers known	

[a] The Stop mutation is coupled with a translocation between chromosomes II and IV, and crosses of Stop with other markers showed whether or not they were linked to chromosomes II and IV, which could be identified cytologically; if unlinked but on an autosome, it could be deduced that the marker was on chromosome III (Crouse and Smith-Stocking 1938). Cytological identification of linkage groups was further defined by Crouse (1943) by genetic crosses using other translocations.

1.2.2 The Germ Line Limited "L" Chromosomes

The chromosome constitution of the germ cells of both males and females is depicted in Fig. 1 C. Three pairs of autosomes are present, and males as well as females both have two X chromosomes. In addition, large heteropycnotic metacentrics are found, which are the germ line "limited" or "L" chromosomes. Typically there are two L chromosomes in primary oocytes or primary spermatocytes of *S. coprophila*, but in 22% of the cases in primary spermatocytes the number of L chromosomes may vary from zero to four, presumably due to mitotic nondisjunction in the six gonial divisions preceding spermatogenesis (Rieffel and Crouse 1966, Crouse et al. 1971). L chromosomes are found in the germ cells of most but not all species of *Sciara*, as will be discussed in greater detail later. Originally Metz thought that L chromosomes were only in the male germ line, since he did not see them in ovarian follicle nuclei, and they were called "male limited" chromosomes as opposed to the four pairs of "ordinary" chromosomes in the soma (Metz 1925 b). Contrary to his earlier belief (Metz 1925 b, Metz and Moses 1926), he deduced that the L chromosomes are not sex-determining chromosomes because he found a species of *Sciara* which lacked any L (Metz 1929 b). Later he discovered that L chromosomes were in the female as well as the male germ line (Metz and Schmuck 1931 a, b, Schmuck and Metz 1932), and this observation confirmed his previous deduction that L chromosomes seem unrelated to sex determination.

1.3 Sex Determination

1.3.1 The X' Chromosome

It is a credit to Metz and his coworkers that they were able to unravel the mode of sex determination in *Sciara*, as it is so unusual. Metz discovered early on that the female in *S. coprophila* determines the sex of her offspring: she will have either all daughters or all sons, but not a brood of mixed sex. For example, if a male is used for more than one mating, offspring of his from one mating may be all sons and offspring of his from another mating may be all daughters (Metz and Moses 1926). Similarly, if one female was placed in a vial with several males, she would have only unisexual progeny; mating one female by a wild-type male who was then subsequently replaced in the vial by a male with the truncate marker (or vice versa) indicated that a female will only mate with the first male she encouters (Moses and Metz 1928). Dr. E. B. Wilson coined the term "monogenic" to describe strains of *Sciara* with unisexual progeny, as contrasted to "digenic" strains of *Sciara* which have bisexual progeny from a single mother (cited in Metz 1931 a, and Metz and Schmuck 1931 a). In monogenic strains, such as *S. coprophila*, a gynogenic mother produces only daughters and an androgenic mother produces only sons.

A female-producing mother breeds as if heterozygous (Aa) and a male-producing mother breeds as if homozygous (aa) for a recessive gene governing sex of progeny (Metz and Moses 1928); the male would be homozygous (aa) as well. It was proposed by Metz and Schmuck (1929 a) that a male-producing mother is

XX while a female-producing mother is X'X, where X' (X-prime) differs from the X by a sex determination agent. If the prime factor is a single allele, one might expect to see crossing-over onto the X which was marked with a recessive mutant (swollen or narrow), but this was never observed (Metz and Schmuck 1929a). Moreover, the dominant marker Wavy on the X' chromosome has never been found to cross-over onto the X (Metz and Smith 1931). There is a very low frequency of crossing-over even between two X chromosomes, with only a frequency of about 0.5% being measured between swollen and narrow (Metz and Schmuck 1931a). Therefore, if the prime factor mapped between swollen and narrow, a cross-over frequency of only 0.25% would be expected, which is hardly measurable (Metz and Schmuck 1931a). Metz and Schmuck (1931a) concluded that if prime is not between swollen and narrow, then either there must be more than one genetic locus for X prime factors, or else there may be only a single prime factor but a lack of synapsis between X and X' and therefore no crossing-over. In *S. coprophila* the X' chromosome carries a long paracentric inversion whose break points have been cytologically mapped (Crouse 1977), and which Crouse (1960a) suggested may prevent recovery of cross-over products with the X. A long paracentric inversion also occurs on the X' chromosome of *S. impatiens* and it was found that primary oocytes had many dicentric and acentric chromatids, due to cross-overs within the inversion between X and X' (Carson 1946). In the second meiotic division, the dicentric remained between the two inner nuclei of the four nuclei formed by meiosis, and the ootid nucleus was derived from one of the terminal nuclei; therefore, directed segregation of the non-cross-over nuclei resulted in a failure to transmit the X-X' cross-over product through the ovum (Carson 1946), confirming the hypothesis of Sturtevant and Beadle (1936) for cross-overs within a paracentric inversion.

1.3.2 Role of the Sperm in Sex Determination

If sex is determined by the female in *Sciara*, what might the role be of the sperm for sex determination? Initially Metz thought that the L chromosomes were limited to the male germ line and involved in sex determination (Metz 1925b, Metz and Moses 1926). However, if this were so then all sperm should be male-determining and all fertilized eggs should become male (Metz 1925b). He further postulated that female offspring should arise either by parthenogenesis (Metz 1925a) or by subsequent chromosome elimination, but the lack of experimental evidence for parthenogenesis spoke against the former idea (Metz 1925b). Within a few years it was necessary to abandon this hypothesis because Metz concluded that the L chromosomes were not involved in sex determination after all, as discussed in Section 1.2.2 (Metz 1929b, Metz and Schmuck 1931a, b, Schmuck and Metz 1932).

The next model for sex determination revolved around the idea that there were X- and Y-bearing sperm produced in equal numbers (Metz and Ullian 1929). In order to accomodate the observation that females of monogenic strains of *Sciara* have only all sons or all daughters (Metz and Moses 1926), Metz proposed that female-producing mothers have eggs fertilized only by X-bearing sperm and male-producing mothers have eggs fertilized only by Y-bearing sperm (Metz

1929 a). If there were differential mortality after fertilization, then 50% of the zygotes should die, but there was no evidence for degenerating fertilized eggs prior to egg-laying, and often there was over 70% viability of eggs once they were laid (Metz 1925 a). The only alternative left was that in female-producing mothers the Y-bearing sperm could not function, and that in male-producing mothers the X-bearing sperm could not function (Metz 1925 a), but the mechanism for this was not spelled out.

There were other problems as well with the selective fertilization model. It was already known, as will be described shortly, that the paternal chromosome set was eliminated at meiosis I of spermatogenesis, so presumably the Y chromosome would also be eliminated in this group (or else one would have to suppose that the X and Y escaped the maternal-paternal imprint and segregated randomly in meiosis I – an idea Metz also considered briefly). In the secondary spermatocyte a "precocious" chromosome presumed to be the sex chromosome (i.e., the maternally derived X) underwent nondisjunction and this dyad was always incorporated in the future sperm. It was necessary for Metz to postulate that pulling by spindle fibers from opposite poles caused the precocious dyad to break and transformed one of the sister X chromatids into a Y (Metz 1930). In this way each sperm would then gain one X and one Y (without such a proposal there seemed no way to obtain Y-bearing sperm). Then selective action by the egg after fertilization would eliminate either one X or one Y brought in by the sperm, depending on the sex of the zygote (Metz 1930). This model seemed more compatible with the existing data than his model of selective fertilization proposed a year earlier.

The reader should realize that up to this point Metz presumed that a Y chromosome must be present because cytological examination of ovarian follicle cells and spermatogonia showed that both had eight ordinary chromosomes, suggesting that the Y was a rod which looked the same as the X (Metz 1925 b). Therefore, imagine everyone's surprise when further cytological study showed that the male soma of *S. coprophila* had only seven ordinary chromosomes unlike the eight present in the male germ line (Metz 1931 b), but the female soma had eight ordinary chromosomes like the female germ line (Metz and Schmuck 1931 b). These observations supported the idea mentioned to Metz by Dr. M. Demerec that the male germ line of *Sciara* might be XX instead of XY (cited in Metz 1931 a, b), and that the mother determines the sex of the zygote by regulating X chromosome

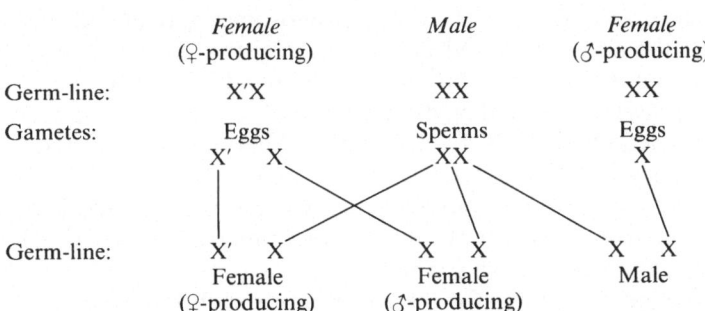

Fig. 2. Diagram of the sex chromosome mechanism for production of unisexual families in *S. coprophila*. (Taken from Metz 1938)

elimination in the embryo. X chromosome elimination would reduce the male soma to a single X (XO), and the male germ line would be XX (Metz 1931 b). Finally the correct model had emerged! This model was further substantiated by direct cytological observation of X chromosome elimination in the embryo. Figure 2 summarizes this correct model for the sex chromosome mechanism for production of unisexual families in *S. coprophila*. The use of the dominant marker Wavy on the X' and/or the recessive markers swollen or petite on the X conveniently allow the researcher to predict the sex of progeny of a female.

1.4 Overview of the Chromosome Cycle

Figure 3 depicts the history of chromosomes during development and gametogenesis in *S. coprophila*. Oogenesis is orthodox and results in a haploid egg with one copy of each ordinary chromosome and generally one L chromosome. This egg is fertilized by a sperm which has a haploid number of autosomes but two sister chromatids of the X, and generally also two L chromosomes. Therefore, in the resulting zygote the haploid number of autosomes is restored, but three X chromosomes (XXX or X'XX) and generally three L chromosomes result. To prevent an accumulation of X and L chromosomes in successive generations, a series of chromosome elimination events occur in the embryo. All L chromosomes are eliminated from somatic cells, and generally all but two L chromosomes are eliminated from the germ line; this occurs identically in male or female embryos. In both male and female germ lines one of the two paternal X chromosomes is eliminated, so that both male and female germ lines now each have two copies of the sex chromosome. The X chromosome elimination in the somatic line differs between male and female embryos. If the mother was androgenic (XX), then all the progeny will be sons and these embryos will eliminate two paternal X chromosomes to achieve the male somatic condition of XO. On the other hand, if the mother was gynogenic (X'X), then all the progeny will be daughters and the embryos will eliminate only a single paternal X to achieve the female somatic condition of XX or X'X.

Spermatogenesis is highly unusual. In the first meiotic division a monopolar spindle forms and the paternal set of ordinary chromosomes (autosomes plus X) "move backwards" away from the single pole to be discarded in a bud of cytoplasm. The maternal set of ordinary chromosomes and all L chromosomes are collected at the single pole and are retained by the spermatocyte. The second meiotic division is apparently bipolar, and all dyads except for the X dyad align on the metaphase plate and then disjoin normally to opposite ends of the spindle. However, at metaphase the X dyad is not found at the equatorial plate but instead is seen "precociously" at one pole. Due to nondisjunction of the X dyad, one product of meiosis II is a chromosome set which is nullo-X and is eliminated in a bud of cytoplasm. The other product has two copies of the X in the chromosome set, and differentiates into the sperm.

The evidence and details for these events in the chromosome cycle of *Sciara* will be described in greater detail in the following sections and mechanisms governing these events will be discussed. Of major interest are the questions:

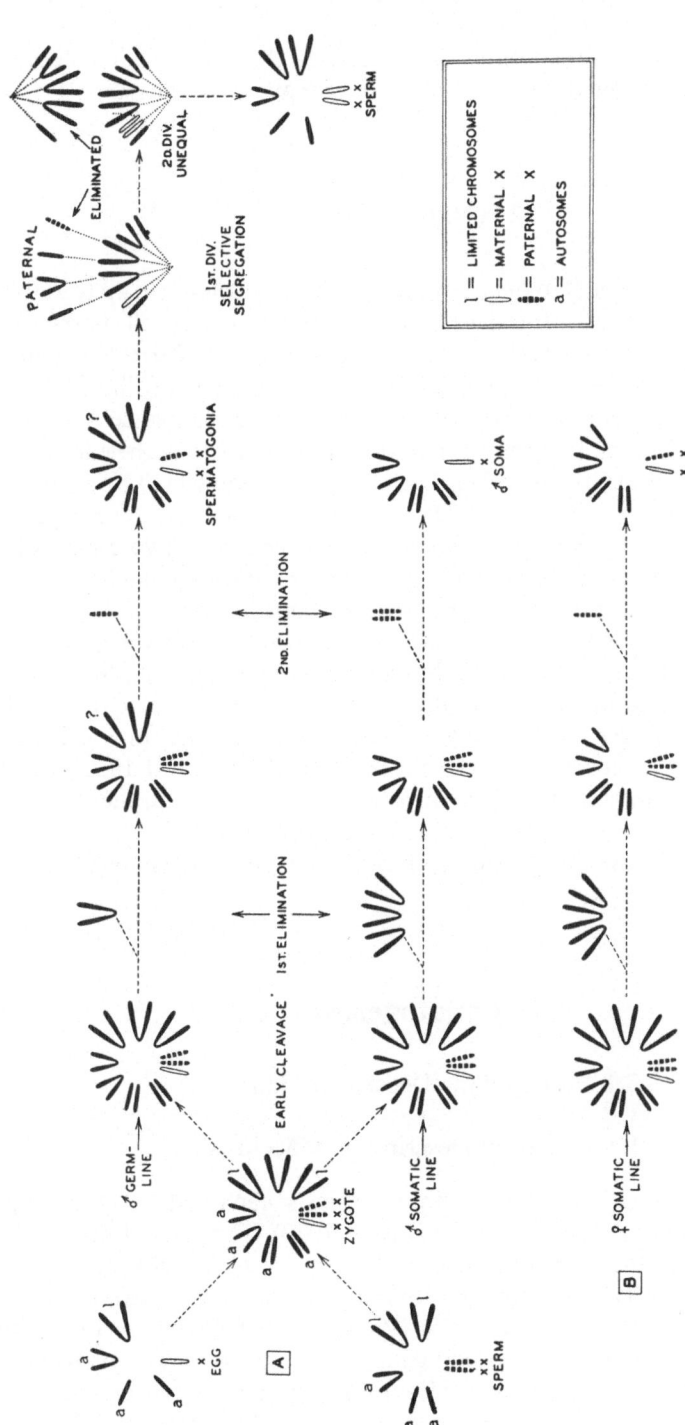

Fig. 3 A–B. Diagram of chromosome eliminations during development and gametogenesis in *S. coprophila*. **A** Chromosome eliminations in the male germ line and male somatic line. The elimination events in the female germ line are the same for the male embryo as those depicted here for the male embryo; oogenesis is orthodox, unlike the unusual events in spermatogenesis. **B** Chromosome eliminations in the female somatic line which differ from the male in that only one instead of both paternal X chromosomes is eliminated. In females, the maternally derived sex chromosome may be either X or X'. (Taken from Metz 1938)

a) How do chromosome eliminations differ between the somatic and germ lines?
b) How do chromosome eliminations differ between the male and female soma?
c) What factors determine differentiation into one sex or the other?

2 Oogenesis

Oogenesis is conventional in *Sciara*. In *Sciara ocellaris* there are up to 200 cells per ovary; on oogonium differentiates into a nurse cell or an egg cell (Berry 1941). Each follicle is believed to have one nurse cell and one oocyte (Berry 1941), unlike the earlier statement by DuBois (1932a) of three nurse cells per follicle (she may have confused degenerating follicles with nurse cells). The nurse cells are polyploid (Berry 1941), but the L chromosomes do not participate in the endomitotic duplications (Himes and Crouse 1961). The nurse cell is connected by a cytoplasmic bridge to the oocyte (Phillips 1967). The primary oocyte has three pairs of autosomes, two copies of the sex chromosome, and generally two copies of the L chromosome (Metz and Schmuck 1931b). There is synapsis of all homologs in prophase of the first meiotic division, and five chromosome pairs can be counted (Metz et al. 1926, Berry 1941). Genetic evidence has also been obtained for synapsis in the female germ line (Metz and Schmuck 1929b). Segregation of homologs in female meiosis is random, as is generally true for most biological systems (Metz 1926b). In some instances there is only a single L in the oocyte, in which case it segregates randomly to one pole or the other (Schmuck and Metz 1932). Oogenesis is arrested at metaphase of the first meiotic division, and continues only after a sperm pronucleus has entered the egg (Schmuck and Metz 1932). A single haploid ootid plus polar bodies result from conventional meiotic divisions of each oocyte.

3 Spermatogenesis

3.1 First Meiotic Division in Males

3.1.1 Cytological Description of Meiosis I

The testis is composed of follicles with many cysts; each cyst is a clone of germinal cells embedded in epithelial cells (Phillips 1970). The cells of one cyst are actually a syncytium due to cytoplasmic bridges between the cells, and the cells within one cyst are fairly synchronous (Phillips 1970). Cysts of greater maturity are found nearer the center of the testis (Phillips 1970). Prophase of the first meiotic division in males and the preceding interphase are periods of intense RNA synthesis, as visualized cytologically by ^{3}H-uridine uptake in *Rhynchosciara angelae* (Basile 1970). Unlike the situation in females, in prophase of meiosis I in males there is no synapsis of homologs (Metz 1925a, Metz et al. 1926). Synapsis of

homologs is not seen in spermatogonia either, even at as early a stage as 16 cells/
testis (Metz 1931 b). Typically one can count ten chromosomes in primary spermatocytes in prophase: six autosomes, two X chromosomes and generally two L
chromosomes (which are more condensed than the "ordinary" chromosomes)
(Metz et al. 1926).

Unlike conventional divisions, there is no true metaphase following prophase
– the chromosomes do not align near the center of the spindle (Metz 1926a, Metz
et al. 1926). Instead, the primary spermatocyte proceeds directly from prophase
to an anaphase-like state in which homologous chromosomes separate from one
another. Figures 4 and 5 depict the stages of meiosis I in spermatocytes. A remarkable aspect of this division is the appearance of a spindle which is monopolar
("monocentric") rather than bipolar (Metz 1925a, 1926a, Metz et al. 1926). This
unusual spindle has recently been studied by electron microscopy (Kubai 1982):
the spindle microtubules do indeed radiate outward from a single pole, and no
centrioles or pole-like structure are seen beyond the broad end of the spindle. At
the acuminate end of the monopolar spindle, a "polar complex" can be visualized;
the polar complex is composed of a ring-like structure, named the "polar organelle", surrounded by a mass of flocculent material (Kubai 1982). A monocentric
division can be artificially induced, for example in injured sea urchin eggs (Wilson
1925), but is a pathological and abnormal situation. However, monocentric division is the normal means utilized for meiosis I in *Sciara* spermatogenesis. A
monopolar spindle has been seen in *S. coprophila* (Metz 1925a, Metz et al. 1926),
S. simulans (Metz 1926a, Metz et al. 1926), *S. prolifica* (Metz et al. 1926), *S. pauciseta* (Metz et al. 1926), *Plastosciara pectiventris* (Fahmy 1949), *Rhynchosciara
angelae* (Basile 1970), *Trichosia pubescens* (Amabis et al. 1979), and all other
Sciarids that have been studied (Metz 1936). The maternal or paternal origin of
the chromosomes on the monopolar spindle, and theories on how chromosomes
move on this unique spindle will be discussed below.

By telophase, the set of four "ordinary" chromosomes that moved away from
the single pole have now been collected together at the periphery of the cell and
are excluded in a bud of cytoplasm (Metz et al. 1926). Each cyst of the testis con-

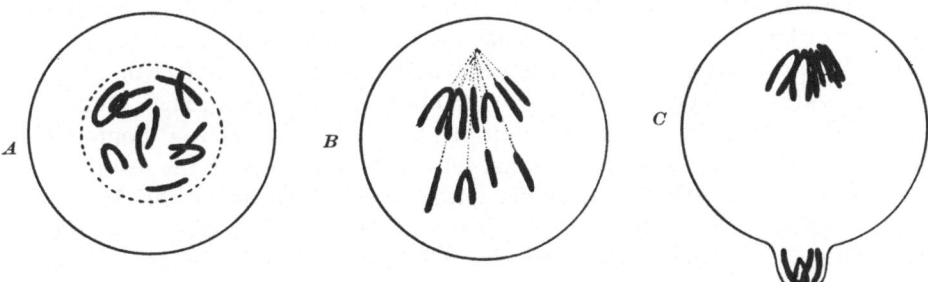

Fig. 4 A–C. Schematic diagram of the first spermatocyte division in *S. coprophila*. **A** Prophase
with ten chromosomes that have not synapsed. **B** Anaphase-like stage on the monopolar spindle.
The two L chromosomes and four "ordinary" chromosomes go towards the single pole, and the
other set of four "ordinary" chromosomes appear to move in the opposite direction, away from
the single pole. **C** Telophase, showing the expulsion of four chromosomes in a bud. Data is from
Metz, as represented schematically by DuBois (1933)

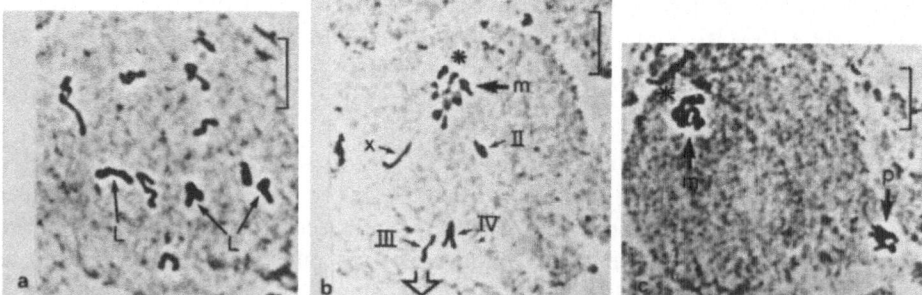

Fig. 5a–c. Meiosis I in *Sciara coprophila* spermatogenesis. **a** Primary spermatocyte in late prophase I. *Arrows* indicate the large heterochromatic supernumerary L chromosomes. The other eight chromosomes represent the maternal and paternal sets. **b** Primary spermatocyte in anaphase I. *m* maternal and supernumerary chromosome set. The other four paternal homologs are migrating to the cell periphery in the direction of the *large open arrow*. Their probable chromosome identification was made on the basis of size and shape. *Asterisk* is the approximate location of the single spindle pole. **c** Primary spermatocyte in telophase I. *p* the paternal bud; *m* the maternal and supernumerary L chromatin; *asterisk* the approximate location of the single spindle pole. Bar = 10 μm. (Taken from Abbott et al. 1981)

tains a group of spermatocytes that are roughly synchronous in meiotic stage, and are arranged in a circle around a central lumen; the spermatocytes have a radial polarity such that each cell forms a bud into the lumen of the cyst (Metz et al. 1926, Kubai 1982). The result of bud formation is that the four excluded chromosomes will not be genetically passed on through the sperm. The bud starts to form at the beginning of meiosis I in *Trichosia pubescens* (Amabis et al. 1979), unlike *Sciara*, where it forms at telophase I.

3.1.2 Genetic Evidence for Segregation of the Paternal Chromosome Set

Originally Metz thought that the L chromosomes were only found in males, so he proposed that the group of four chromosomes that went together with the two L chromosomes to the single pole might be all paternally derived, whereas the four ordinary chromosomes excluded in the bud would be maternal in origin (Metz 1925 a). As discussed above (Sect. 1.2.2), it later became clear that L chromosomes are in females as well as males. However, subsequent genetic studies did indeed show that there is segregation of maternal from paternal homologs in meiosis I, but in fact it is the maternal set which is retained in the future sperm and the paternal set which is discarded in the bud (opposite from what Metz originally proposed). Note that such directed segregation of homologs is unique; in most other biological systems there is random segregation at meiosis I such that any one pole may receive, for example, maternal homolog A, paternal homolog B, paternal homolog C, maternal homolog D, etc. The genetic studies which proved that the paternal set of homologs is discarded in the bud utilized the marker truncate on chromosome II (Metz 1926 b, Metz 1927), and subsequently utilized a variety of other markers on the other chromosomes of *S. coprophila* (Smith-Stocking 1936). Selective segregation to discard paternal homologs in meiosis I was also proven by genetically marked crosses in *S. simulans* (Metz 1928), *S. impatiens* (Metz 1929c), and *S. ocellaris* and *S. reynoldsi* (Crouse and Smith-Stocking 1938). It is clear from cytological examination that the L chromosomes do not segregate selectively – they are all always collected at the single pole and not eliminated in the bud (footnote 2 of Metz et al. 1926, Metz 1938). This was further substantiated by crossing an L-minus strain with an L-plus strain of *Sciara impatiens*: regardless of paternal or maternal origin of the L chromosomes, the L chromosomes were never discarded in the bud resulting from meiosis I of spermatogenesis (Crouse et al. 1971).

3.1.3 Mechanism of Chromosome "Imprinting"

How does the primary spermatocyte recognize the paternal chromosomes to be eliminated and distinguish them from the maternal chromosomes to be retained? In other words, what is the "imprint" on the paternal chromosomes and/ or the "imprint" on the maternal chromosomes? Finally, how do the L chromosomes avoid the maternal-paternal "imprint"? Chandra and Brown (1975) defined "imprinting" as the differential behavior between homologous chromosomes that has been predetermined several cell divisions before the actual manifestation of the differential behavior. Rieffel and Crouse (1966) supposed that both maternal and paternal sets of ordinary chromosomes carried an imprint, and only the L chromosomes escaped being imprinted. The imprint must be reversible because the maternal chromosomes set that is retained by the sperm is then recognized as a paternal set once the sperm has fertilized the egg in the next generation. Rieffel and Crouse (1966) further speculated that the paternal imprint was removed in the embryonic germ line during a stage when chromosomes of only the maternal set are diffuse in appearance. Later, Chandra and Brown (1975) suggested that it would be only necessary to imprint one of the two sets of homologs to fit the observations. In the male zygote of the mealy bug the paternal homologs are inactivated by heterochromatization, and they proposed that the paternal set is imprinted in the fertilized egg before fusion of the pronuclei (Chandra and Brown 1975, Brown and Chandra 1977). It is unknown whether one or both sets of homologs is imprinted in *Sciara*, and lack of data allows for numerous models.

There is no data on the time or mechanism for chromosome imprinting in *Sciara*. It has been postulated that methylation of DNA may play a role in the imprinting process (Sager and Kitchen 1975). In this model the maternal chromosomes would be methylated and the paternal chromosomes would not be methylated, rendering the latter susceptible to a complex restriction-modification system (Sager and Kitchen 1975). The need for a restriction-modification system could be obviated, if methylation altered DNA-protein interactions (e.g., at the kinetochore). The scanty evidence so far is not in support of imprinting via methylation. First of all, in general, vertebrate sperm DNA is methylated (reviewed by Bird 1984), contrary to the Sager and Kitchen model (1975) that the paternal chromosomes are not methylated. Secondly, methylation of DNA probably does not exist in *Drosophila* and other insects, and in fact is rare in invertebrates in general (Bird 1984). The only controversy on this point lies with the observation that salivary gland polytene chromosomes but not brain chromosomes of *Drosophila* and *Sciara* seem to bind antibodies against 5-methyl cytosine (Eastman et al. 1980, Wei et al. 1981).

Studies in my own laboratory to search for methylation in *Sciara* testis DNA have been negative. Antibody against 5-methyl cytosine failed to bind to meiotic chromosomes of *Sciara coprophila* spermatocytes (S. A. Gerbi, D. A. Miller, O. J. Miller and B. F. Erlanger unpublished observations). In addition we tested cleavage by the restriction enzyme isoschizomers Msp I and Hpa II; unmethylated DNA can be cut by both enzymes, but when the internal cytosine of the restriction site (CCGG) is methylated then only Msp I and not Hpa II will cut it. Genomic DNA isolated from 20 testes dissected from pre-meiotic (eye filled 1/3 with pigment), meiosis I (eye filled 1/2 with pigment), or post-meiotic (eye filled 5/6 with pigment) stage *Sciara coprophila* pupae showed the same restriction bands as cloned (and therefore unmethylated) *Sciara coprophila* ribosomal DNA (pBC2) in a Southern blot hybridized to ^{32}P-pBC2 DNA (K. Howard and S. A. Gerbi unpublished observations). The same restriction pattern was also seen in the controls of DNA from male larval brains, male larval salivary glands, and larval ovaries. Therefore, there is no biochemical evidence for DNA methylation in the germ line of *Sciara* males or females, or in diploid or polytene male somatic cells. It is still formally possible, although unlikely, that methylation occurs elsewhere than at the Hpa II sequence or elsewhere in the genome besides the ribosomal DNA cluster.

To sum up, it seems implausible that DNA methylation is part of the imprinting mechanism. How and when the chromosome sets of *Sciara* become imprinted remains an elusive mystery.

3.1.4 Mechanism for Chromosome Movement on the Monopolar Spindle

Metz believed that all chromosomes in the primary spermatocyte are attached by spindle fibers to the single pole (Metz 1926a, Metz et al. 1926). However, recent electron microscopic observations demonstrate that this is not the case (Kubai 1982). The L chromosomes seem to lack kinetochores and spindle fibers, the maternal homologs each have a kinetochore but are not attached to spindle fibers, and it is only the paternal chromosomes that are attached to spindle fibers via their kinetochores (Kubai 1982). These recent data also show that the L chromosomes never really move and are always adjacent to the single pole, even in early primary spermatocytes. Metz thought that the maternal homologs moved in a normal fashion to the single pole; however the ultrastructural observations of Kubai (1982) suggest that the maternal homologs might not even be oriented in the usual fashion. Instead, the kinetochore on each maternal homolog is not necessarily pointed towards the monopole (Kubai 1982). How the maternal chromosomes move to the pole without spindle fiber attachments is an unexpected dilemma and current problem.

Most research has focused attention on the unique movement of the paternal chromosomes away from the single pole. By cytological inspection it was clear that chromosome IV, which is a metacentric, moves "backwards" because its centromere lags behind rather than leading the direction of movement (Metz 1926a, Metz et al. 1926). Metz presumed that therefore all the paternal chromosomes move "backwards"; we have recently confirmed that this is indeed the case for the three acrocentrics (II, III, X) as well as the metacentric (IV) by DAPI c-banding and in situ hybridization (Abbott et al. 1981). However, the set of chromosomes to be excluded in the bud of the primary spermatocyte of *Trichosia pubescens* lacks a characteristic shape (Amabis et al. 1979), unlike the backwards orientation seen for paternal chromosomes in *Sciara*. Metz observed that the lagging end of each paternal chromosome is taut, suggesting spindle fiber tension holding it back (Metz 1926a); in modern terminology we could say that the kinetochore microtubules act as governors to restrain the force pulling the paternal chromosomes away from the monopole. Kubai (1982) has suggested that the kinetochore microtubules prevent the paternal chromosomes from moving to the monopole, while the maternal chromosomes, which are unattached to spindle fibers, readily move to the monopole.

What is the force that pulls the paternal chromosomes away from the single pole? Metz (1926a, 1933) saw no evidence for traction on the leading end of the paternal chromosomes, and we have confirmed by X-ray breakage that no neocentromeric force seems to tug at the free chromosome ends despite the telomeric location of satellite DNA (Abbott et al. 1981). Metz interpreted the unique chromosome movement in the primary spermatocyte as due to differing electrical charge of the two sets of chromosomes (Metz 1926a), and he discussed how this unique system in *Sciara* did not fit any of the models for standard chromosome movement (Metz 1933). He concluded that the paternal chromosomes seem to move with an autonomous force (Metz et al. 1926, Metz 1933), since each paternal chromosome follows its own path that is apparently determined only by its random original location at the start of meiosis I (Metz 1933). The argument for autonomous chromosome activity continued that each paternal chromosome is always at the apex of a pseudopod of gel (nucleoplasm?); if the paternal chromosome were just moved only by flow of the surrounding gel, then the retarding force of the spindle fiber attached to it should cause the chromosome to move more slowly than the other material in the gel and not faster (Metz 1933). Electron microscopy has shown that the nuclear envelope remains fairly intact during meiosis I (Kubai 1982), and the nuclear envelope blebs out around each paternal chromosome (i.e., the gel pseudopods of Metz). A basket-like mesh of microtubules lines the inner side of the nuclear membrane, but is not directly apposed to the paternal chromosomes, thereby obscuring its potential role in chromosome movement.

No buffered medium has yet been found to successfully support the culture of living spermatocytes from *Sciara*, so in vitro studies with inhibitors have not been possible to elucidate what the force is for chromosome movement. However, decades ago Metz did try one experimental treatment in which male pupae were placed in the cold to block paternal chromosome movement and bud formation; nonetheless, the paternal dyads divided into sister chromatids while still artificially kept in the primary spermatocyte (Metz 1933). Metz concluded that the paternal chromosomes were still "alive", and not already degenerating in meiosis I.

Although isolated *Sciara* spermatocytes have not been kept successfully in a saline solution, it has been possible to maintain them alive for several hours under oil (S. Gerbi and R. B. Nicklas unpublished observations). Unfortunately, time lapse cinematography is hampered by the mitochondria, which appear as large refractile globules (Metz et al. 1926) and obscure most of the monopolar spindle from view (S. Gerbi and E. D. Salmon unpublished observations, Zilz 1970).

Kubai (1982) has speculated that perhaps the paternal chromosomes never really move, as the monopolar spindle length does not increase and the distance of paternal chromosomes from the single pole seems to remain constant during anaphase I; she continues that the paternal set may already be segregated from the other chromosomes by prophase I. It is clear that the paternal chromosomes must have their position altered by telophase I when they are enclosed in a bud for expulsion from the spermatocyte, but the nature of this force remains unknown.

Therefore many unsolved problems still must be unraveled before our understanding is complete for meiosis I of *Sciara* spermatogenesis. The mechanism of chromosome movement still needs to be explored further. How do the maternal chromosomes move towards the monopole if they lack kinetochore microtubules? Do the paternal chromosomes move "backwards" away from the single pole; if, on the other hand, they do not move during anaphase I, how are they captured into the bud for expulsion from the primary spermatocyte? We are still totally ignorant on the mechanism and timing of imprinting paternal vs. maternal chromosome sets. How does the imprinting mechanism effect the failure of spindle microtubules to attach to the maternal kinetochores? If the imprint is present prior to spermatogenesis, how can the maternal chromosomes have kinetochore microtubules in premeiotic divisions in the germ line?

3.2 Second Meiotic Division in Males

3.2.1 Cytological Description of Meiosis II

A bipolar spindle forms in the second meiotic division in male *Sciara* (Figs. 6 and 7). Generally six chromosomes (four ordinary and about two L chromosomes) remain in the spermatocyte, having been gathered together at the monopole of meiosis I. By metaphase of meiosis II, five of these six dyads align themselves on the equatorial plate of the spindle. However, one dyad is seen "precociously" at one pole of the spindle (Metz 1925a), where it will later divide into two sister chromatids (Metz 1925b). At anaphase II the sister chromatids of the dyads on the equatorial plate separate and move to opposite poles. However, one group of separated chromatids will lack a copy of the precocious chromosome, and this group will be expelled from the secondary spermatocyte in a bud of cytoplasm to degenerate. The nucleus which remains in the secondary spermatocyte has two identical sister chromatids of the precocious chromosome.

Metz (1930) guessed correctly that the precocious chromosome was the sex chromosome. This view was supported by Schmuck (1934), who observed that the precocious chromosome of meiosis II had the same size and shape as the two iden-

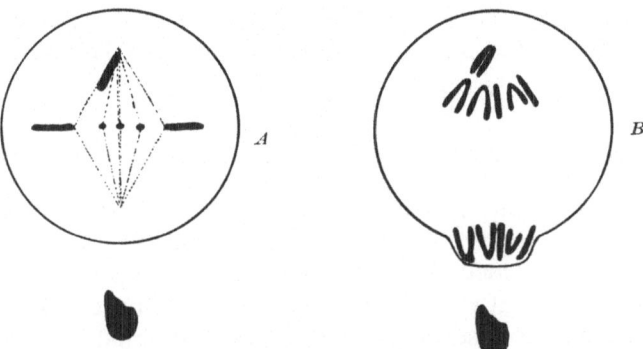

Fig. 6 A–B. Schematic diagrams of the second spermatocyte division. **A** Metaphase II showing the precocious dyad already at one pole, while the other chromosomes are still aligned on the equatorial plate. **B** Telophase II showing the formation of a bud with a nullo-X chromosome complement. The bud formed by meiosis I is shown outside the secondary spermatocyte in **A** and **B**. (Taken from DuBois 1933)

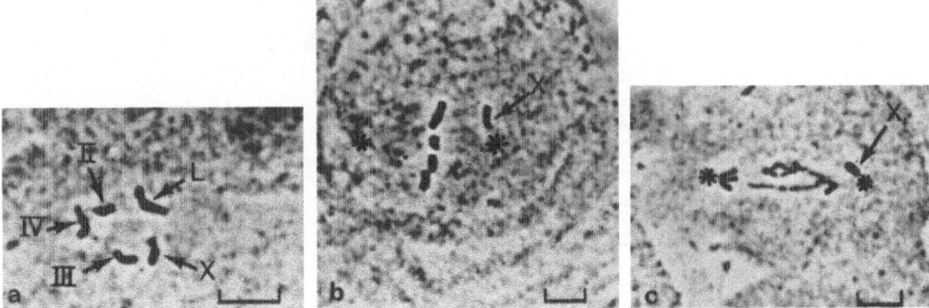

Fig. 7 a–c. The second meiotic division of *Sciara coprophila* spermatogenesis. **a** Prophase secondary spermatocyte showing the four maternal homologs (dyads) and a single L dyad. Chromosome assignments are determined by size and shape in this and other photographs. **b** Metaphase secondary spermatocyte illustrating that the precocious X-dyad is not on the metaphase plate but found adjacent to one of the spindle poles. *Asterisks* mark the approximate location of the spindle poles. **c** Anaphase secondary spermatocyte showing the clearly defined bipolar nature to the spindle. Those dyads which were previously on the metaphase plate have divided while the X dyad (juxtaposed to one pole) does not divide. Bar = 10 μm. (Taken from Abbott and Gerbi 1981)

tical chromosomes later eliminated from the male soma (a rod-shaped acrocentric in *S. coprophila* and a V-shaped metacentric in *S. pauciseta*). Since the male soma is XO in constitution after this embryonic elimination event, unlike the female soma, which is XX having eliminated only one chromosome, the eliminated chromosome must be the X. By deduction, the precocious chromosome of meiosis II with the same morphology must also be the X. This logical conclusion was also echoed by Metz (1934). In *S. reynoldsi* the four ordinary chromosomes are morphologically distinguishable, and it was possible to confirm that the X is the precocious chromosome of meiosis II and the chromosome eliminated differentially from the embryonic soma (Crouse 1943). A series of reciprocal translocations between the X and an autosome showed that this was also the case for *S. coprophila* (Crouse 1943).

3.2.2 Identification of the Control Locus for Precocious Chromosome Behavior

The X translocations demonstrated that an intact X was not needed for precocious behavior (Crouse 1943). Much of the X could be translocated to an autosome, and the control locus for precocious behavior seemed to reside somewhere on the X in the general vicinity of X centromere (Crouse 1943). Later, Crouse (1960b) extended this observation with several additional X translocations, and succeeded in obtaining a translocation (Tl) which separated the centromere from the proximal heterochromatin on the X. The X centromere and the bulk of the X chromosome with an additional bit of chromosome II translocated onto it was

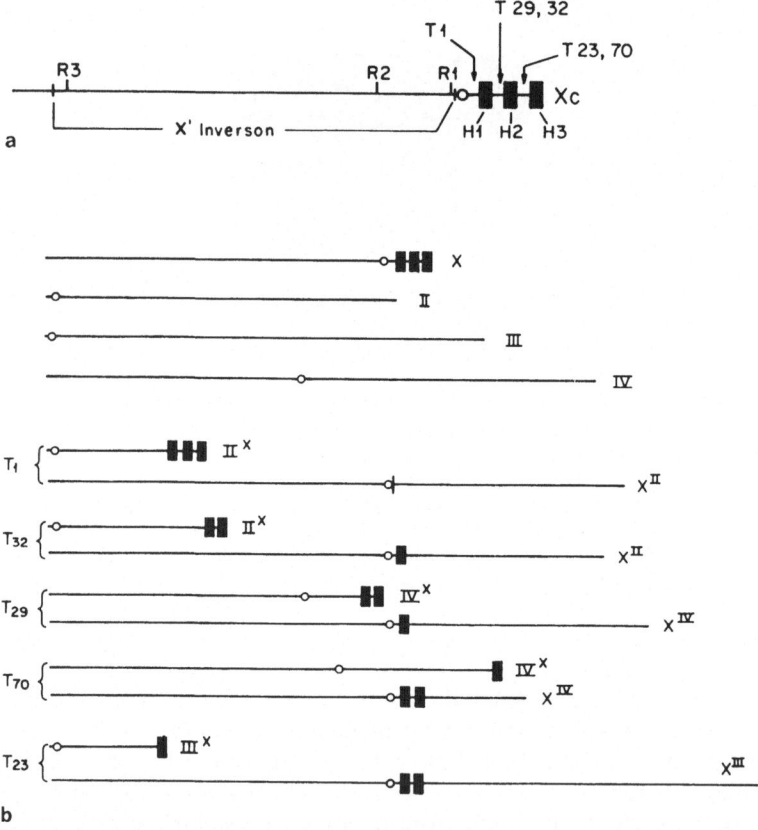

Fig. 8. a Diagram of the X chromosome and three autosomes of *Sciara coprophila*. The centromeres are depicted by *open circles*. The three heterochromomeres in the proximal heterochromatin of the X are shown by darkened rectangles (*H1, H2, H3*), and are separable by the breakpoints of the translocations shown. **b** Resulting products from five reciprocal translocations. Note the new sizes and shapes of these translocation chromosomes. (Taken from Crouse 1979)

seen to lack precocious behavior; instead it went to the metaphase plate in meiosis II and its division products subsequently went to each pole (Crouse 1960b). On the other hand, the tiny bit of the X chromosome contained in the X proximal heterochromatin was now attached to part of chromosome II, and this translocation chromosome now exhibited precocious behavior (Crouse 1960b). Therefore the control locus resides in the X proximal heterochromatin and not at the X centromere. Moreover, the control locus has the ability to control the centromere of an autosome (chromosome II in this case) so that nondisjunction and precocious behavior occurs. Finally, the control locus does not have to be adjacent to the centromere of the chromosome it controls (as is the case in a normal X), because translocation chromosome II^X has the centromere of II at one end and the X proximal heterochromatin at the other end of the chromosome (Fig. 8).

Additional X translocations mapped the X centromere to a specific band in larval salivary gland polytene chromosomes (no chromocenter exists in *Sciara*) (Crouse 1977), and also subdivided the X proximal heterochromatin into three blocks or "heterochromomeres" (Crouse 1979). Figure 8 shows the products of reciprocal translocations involving varying numbers of heterochromomeres from the X proximal heterochromatin. It is possible to identify the translocation chromosomes in meiosis II by their morphology (Fig. 9). Such analysis showed that the control locus resides in the middle heterochromomere of the X proximal heterochromatin (Crouse 1979). These translocations indicated that the control locus is able to govern the behavior of the centromere of *each* of the autosomes (though with more difficulty for chromosome IV in translocation T29) (Crouse 1979). In the X translocation stocks, the embryonic soma eliminates the same chromosome which exhibited precocious behavior in meiosis II, so that the embryo retains a balanced chromosome set (Crouse 1943, 1960b).

In situ hybridization to polytene chromosomes from these X translocation stocks showed that ribosomal DNA (rDNA) is contained in all three heterochro-

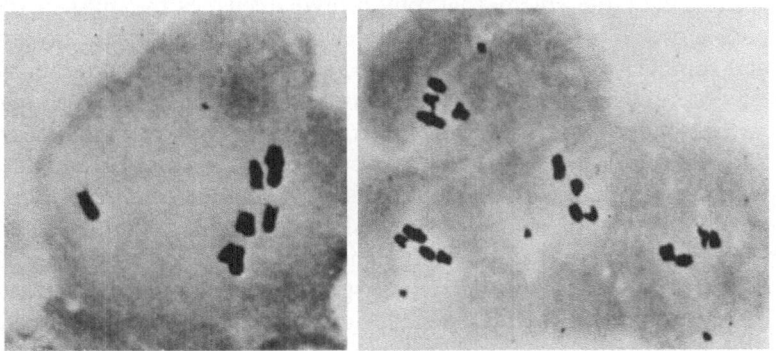

Fig. 9. A Photomicrograph of metaphase II in spermatogenesis of wild-type *Sciara coprophila*. The precocious X chromosome is already at one pole, while the other chromosomes are still on the metaphase plate. **B** Photomicrograph of several spermatocytes in metaphase II of *Sciara coprophila* translocation stock T1. Note that precocious chromosome II^X, which has the centromere of II but three X heterochromomeres at the opposite end of the chromosome, is much smaller than the wild-type X (cf. **A**) or the reciprocal translocation product X^{II} that is on the metaphase plate. (Taken from Crouse 1979)

momeres (Crouse et al. 1977). Thus the control locus must be embedded within the tandem array of rDNA repeat units. An alternate but less appealing hypothesis is that the translocation chromosome with precocious behavior is the one that has the majority of the rDNA repeat units (Abbott and Gerbi 1981).

3.2.3 Mechanism for Precocious Chromosome Behavior

In order to understand how the control locus on the X governs precocious chromosome behavior, it is necessary to distinguish between the following two hypotheses:

1. The centromere of the precocious chromosome is on a different timetable from the other chromosomes, so that the precocious chromosome moves earlier (i.e., at metaphase II) to one pole than the other chromosomes, which do not move until anaphase II. In this scenario, the control locus would govern the timetable for chromosome movement.
2. The centromere of the precocious chromosome is inhibited from normal function (i.e., kinetochore formation and attachment of kinetochore microtubules). The precocious chromosome never moves relative to the pole in meiosis II, but is always already next to the pole. In this case, the control locus would govern inhibition of the centromere of the precocious chromosome.

Cytological observations led Metz to espouse model (1); he believed that the X moved precociously away from the equatorial plate to one pole (Metz et al. 1926). Furthermore, he thought that spindle fibers were attached to the X dyad and went to both poles (Metz et al. 1926). Nondisjunction occurred and both chromatids of the dyad went to the same pole, but tension by the spindle fibers attached to the opposite pole sometimes caused deformation of the precocious dyad (Metz 1934).

Secondary spermatocytes are somewhat more amenable than primary spermatocytes to time lapse cinematography of living cells, as the mass of refractile mitochondria now fills only somewhat under half of the cell volume and most of the bipolar spindle is visible. Metz used fixed and stained material rather than living material upon which to draw his conclusions; we have used living material to analyze which of the two hypotheses described above is correct (S. A. Gerbi and E. D. Salmon unpublished observations, also Zilz 1970). Testes from male pupae with eye anlage filled 2/3 with pigment (meiosis II stage) were dissected under oil and viewed with a combination of Nomarski and polarized light optics. At very early metaphase II, a short stubby spindle is formed and the precocious X chromosome is already next to one pole of this spindle. This spindle elongates considerably during metaphase II, and the precocious chromosome is pulled by the pole away from the equatorial plate. Therefore the distance between the precocious chromosome and the equatorial plate does indeed increase during metaphase II, but the distance does not decrease between the precocious chromosome and the pole because it is always adjacent to the pole. Therefore we favor hypothesis (2), that the control locus inhibits normal centromere function of the X during meiosis II. Electron microscopy will show if the X lacks a kinetochore and kinetochore fibers at this stage (D. F. Kubai personal communication).

The precocious X seems to be anchored to the pole even in early meiosis II; what is the nature of the glue holding it at the pole? We were unable to detect ^{3}H-uridine uptake during metaphase II (Abbott and Gerbi 1981), which rules out massive transcriptional activity of the control locus at this stage – but a low level of activity might not be detected. Is the anchor between the control locus and the pole, or between the centromere of the precocious chromosome and the pole? Normally it would be difficult to address this question because on the wild-type X the control locus maps quite close to the X centromere. However, in translocations T1 and T32 the translocation chromosome II^X has its centromere at one end and the control locus at the other end. These translocation stocks were used for DAPI c-banding and also for rDNA in situ hybridization, and showed that the centromere end rather than the control locus end of translocation chromosome II^X was next to the pole (Abbott and Gerbi 1981). It would appear that the centromere is glued to the pole and inactivated from normal function. How the control locus causes this is unknown, but it has to work in cis rather than trans because it controls only the centromere of the chromosome on which it physically resides. Thus in translocations T1 and T32 the control locus governs the centromere of chromosome II to which it is now linked, and no longer governs the centromere of the X. Moreover, the control locus does not have to be adjacent to the centromere to govern it, as in translocations T1 and T32 it is at the opposite end of the chromosome from the centromere. Is a long-distance effect on chromatin structure at work here?

In our working model we imagine that the X moves to the monopole in anaphase I (in *Plastosciara pectiventris* the X may move somewhat precociously to the monopole in meiosis I; Fahmy 1949) and becomes locked down there even prior to the onset of meiosis II. If this is the case, one would predict that the pole with which the precocious chromosome is associated in meiosis II is a specific and not a random one. Our observations on naturally occurring tetraploid cells are in accord with this prediction: the two X dyads in these cells are both always at the same pole, and we never saw one X dyad at one pole and the other X dyad at the opposite pole (Abbott and Gerbi 1981). We have also seen by time lapse cinematography that the bipolar spindle in meiosis II is asymmetrical: the half spindle with the precocious chromosome forms earlier and is larger and more birefringent than the other half-spindle (S.A. Gerbi and E.D. Salmon unpublished observations). Metz had also observed that the half-spindle near the cell periphery where the bud will form is shorter and less conspicuous than the half-spindle with the precocious chromosome (Metz et al. 1926). Therefore vestiges of a monopolar spindle remain in the bipolar spindle of meiosis II.

In order to delve into the molecular basis for control locus action, we are currently analyzing cloned neighbor sequences of rDNA to identify candidates for the control locus (A. Whitney Kerrebrock, R. Srivastava and S.A. Gerbi).

3.3 Spermiogenesis

Study of spermiogenesis shows that organelle morphogenesis differs between the germ line and soma. The earlier light microscopy observations on spermiogenesis (Doyle 1933) have now been elaborated upon at the ultrastructural level by Phillips and Makielski.

3.3.1 Formation of the Giant Centriole

The centrioles in somatic cells of *S. coprophila* are composed of nine doublet microtubules, rather than the more common pattern of nine triplets (Phillips 1966a). Examination of testes from young male larval (second instar) revealed some nine-membered centrioles but also the appearance of giant centrioles (Phillips 1967). These giant centrioles consist of 20–50 singlet microtubules arranged in an oval; the secondary centriole might be even larger and after further spermatogonial divisions the giant centrioles of the germ line will contain 60–90 singlet microtubules in an oval (Phillips 1967). The centrioles of the female germ line have the same form and developmental pattern as in the male (Phillips 1967).

There are two pairs of giant centrioles in spermatogonial cells; in each pair the primary centriole forms the top of the T-shaped complex, and the secondary centriole is perpendicular to it, intersecting the T (Phillips 1967). In some second instar testes it could be observed that the primary centriole had nine members, while the secondary centriole associated with it was a small giant centriole; the conclusion is that the primary centriole was the mother which gave rise to the daughter secondary centriole (Phillips 1967). Since the mature male and female germ cells have only giant centrioles, and the soma of the embryo has only nine-membered centrioles, it follows that the nine-membered centrioles probably can arise from giant centrioles (or else must arise de novo) (Phillips 1967). Therefore the morphogenesis of centrioles seems to be reversible: giant centrioles can arise from nine-membered centrioles and conversely nine-membered centrioles may arise from giant centrioles.

In spermatogonial divisions the nuclear membrane remains almost intact, and the spindle fibers are not confluent with the centriole pair located outside the nuclear membrane at each end of the spindle (Phillips 1966a, 1967). In the first meiotic division of spermatogenesis, the polar organelle is seen at the apex of the monopolar spindle; it is not clear if the polar organelle is somehow related to a giant centriole (Kubai 1982). By the end of meiosis II in males, a single giant centriole is seen at the pole that is associated with the precocious chromosome, and this giant centriole then migrates caudad to form the base of the flagellum (Phillips 1966a). In the flagellum of the sperm there are 70 doublet microtubules, each with an associated singlet microtubule, thereby reflecting the pattern of 70 singlet tubules in the giant centriole at the flagellum base (Phillips 1966a, b, Makielski 1966).

3.3.2 Mitochondrial Fusion

By light microscopy Doyle (1933) observed that the large spherical mitochondria of young spermatids arrange themselves individually along the spermatid flagellum and later fuse into a giant mitochondrial derivative (the Nebenkern of Retzius 1904). Still later 7–9 h after the sperm is deposited into the spermatheca of the female, much of the contents of the giant mitochondrion is shed (Phillips

1966a, Makielski 1966). Shedding of part of the mitochondrial contents is coupled with acquisition of motility of the sperm, reminiscent in some ways of sperm capacitation in mammals (Makielski 1966). The triggers are unknown for mitochondrial fusion in the spermatid and the subsequent expulsion of some of its contents in the spermatheca.

4 Elimination of Chromosomes from the Soma of the Embryo

4.1 Fertilization and Early Cleavage Divisions

A description of events in early embryogenesis was facilitated by introduction of a method to wholemount the egg and stain it with Feulgen dye, thereby removing the need for sectioned material (Schmuck and Metz 1931). The egg of *Sciara* has a very thin chorion, and under the chorion is a thin yolk membrane (DuBois 1932a). It is impossible to distinguish ventral from dorsal, but the anterior end is marked by the micropyle which is a brownish disc (DuBois 1932a). Presumably the sperm enters the egg through the micropyle, during metaphase I of oogenesis (Schmuck and Metz 1932). After oogenesis is completed, the female pronucleus goes from the periphery to the center of the egg to join the male pronucleus. The male pronucleus donates 6–7 chromosomes (3 autosomes, 2 identical X chromosomes and 1–2 L chromosomes) (Metz et al. 1926), and the female pronucleus donates 5 chromosomes (3 autosomes, 1 X or X' chromosome, and 1 L chromosome) (Schmuck and Metz 1932); therefore, the nucleus of the zygote has 11–12 chromosomes. Chromosome eliminations will reduce this number to 7 chromosomes in the male soma (XO), 8 chromosomes in the female soma (XX or X'X), and 10 chromosomes in the germ line of both sexes. These differences in chromosomal constitution between the soma and germ line led to the theoretical prediction that chromosome elimination must occur in the embryo, before the actual observation was made of such elimination (Metz 1931b, Metz and Schmuck 1931b, Schmuck and Metz 1932). Elimination of entire chromosomes in *Sciara* differs from other systems like *Ascaris* or *Miastor*, where only parts of chromosomes are eliminated (reviewed by Metz 1931b).

4.2 Elimination of the L Chromosomes from the Soma

By the fifth (sometimes sixth) cleavage division, the nuclei have almost reached the external layer of clear cytoplasm, and 1–2 nuclei at the posterior will become the primordial germ cells, thereby effecting the separation of somatic and germ lines (DuBois 1933). There may be 2, 3, or 4 primordial germ cells after the fifth division, depending on the location in the germinative cytoplasm of the daugther nuclei (DuBois 1933). At the fifth (sometimes sixth) division of both sexes, the L chromosomes are eliminated from somatic nuclei, and so the L chromosomes are henceforth limited to the germ line (DuBois 1933). Figure 10 shows L chromosome elimination: the L chromosomes remain either unsplit or split at

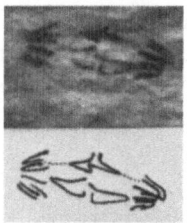

Fig. 10. Sixth cleavage division anaphase of *Sciara coprophila* showing two L chromosomes that have divided but remain at the center of the spindle. The other chromosomes have almost reached the poles; some of the chromosomes at the poles are not visible because they are in a different focal plane. (Taken from Metz 1938)

the center of the spindle in anaphase when the ordinary chromosomes move to the poles (DuBois 1932 b, 1933, Metz 1938). Elimination of L chromosomes is delayed until the sixth division if the nucleus is not yet at the periphery of the embryo (DuBois 1933); this observation supports the idea of Boveri (1904) that chromatin elimination is controlled by cytoplasmic conditions. This conclusion was questioned by Metz (1957) when he saw somatic chromosome eliminations in the central part of the egg as well as in the clear periphery.

What factors in the cytoplasm govern L chromosome elimination, and are these factors absent at the posterior where the germ line is laid down? Do these factors inhibit centromere function of the L chromosomes to prevent poleward movement? What features identify the L chromosomes as the target for inhibition by these factors? Clearly there is much still to be learned about L chromosome elimination. The eliminated L chromosomes remain in the periphery of the yolk as clumps of chromatin (DuBois 1933). Metz (1938) claimed that for 1–2 subsequent divisions the L chromosomes continued to divide in the cytoplasm.

Mistakes can occur in somatic chromosome elimination. DuBois (1933) found deficient chromosome groups in early cleavage stages, and later deficient chromosome groups were also found in the gonads (Metz and Schmuck Armstrong 1961). In ovaries there were deficient cells with only 3–7 (usually 4–5) chromosomes; clusters of cells with the same deficient number of chromosomes indicated that they all derived from the same cell lineage (Metz and Schmuck Armstrong 1961). Nondisjunction is not a likely explanation for the origin of these deficient cells, because cells with a greater than usual number of chromosomes are not found (Metz and Schmuck Armstrong 1961). Also the number of deficient cells per cluster supports the idea that chromosome loss occurred before the gonads had formed (Metz and Schmuck Armstrong 1961). Since several chromosomes besides the L's may be missing from these deficient nuclei, it seems probable that the signals for L and X chromosome elimination in the embryonic soma went awry, and also eliminated autosomes in these deficient cells (Metz and Schmuck Armstrong 1961).

4.3 Elimination of the X Chromosome(s) from the Soma

After elimination of the L chromosomes from the soma, 9 chromosomes remain (6 autosomes and 3 sex chromosomes). In the seventh (sometimes eighth or

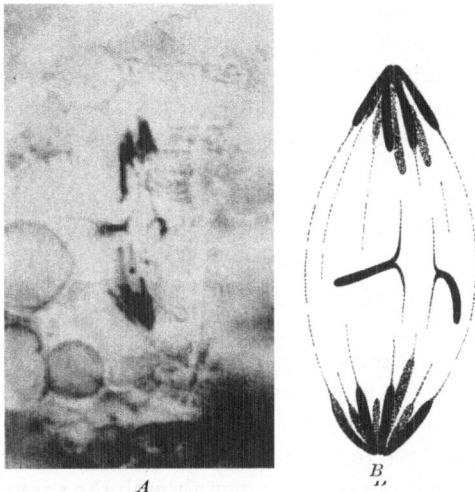

Fig. 11. Seventh cleavage division anaphase of a male embryo of *Sciara coprophila* showing the elimination of two X chromosomes, which are seen to remain undivided at the equatorial plate. The other chromosomes have already reached the poles, and each of the nuclei from this division will be XO in chromosome constitution. B: Schematic presentation of A. (Taken from DuBois 1933)

ninth) division, 1–2 X chromosomes will be eliminated (DuBois 1932 b, DuBois 1933). Like the L chromosomes, the X chromosomes are eliminated by remaining on the equatorial plate during anaphase (Fig. 11) (DuBois 1933). However, unlike L chromosome elimination, the elimination of X chromosomes differs between the two sexes: one X is eliminated from the soma of females and two X's are eliminated from the soma of males (DuBois 1933). The eliminated X chromosomes remain in the periphery of the yolk as clumps of chromatin (DuBois 1933).

The sex of the embryo was predetermined by the mother; presumably the cytoplasm of the egg is conditioned by a factor(s) that governs the differential elimination of X chromosomes between the soma of males and females. How only one X is eliminated in the germ line of males while two X's are eliminated from its soma is a difficult riddle. A search to identify the protein (?) factor(s) that govern X elimination in the soma is underway, using a microinjection assay to reverse the sex of the embryo (W. Büsen, personal communication).

Genetic observations demonstrated that the X chromosome(s) that are eliminated from the soma are always paternal in origin (Metz 1938). Furthermore, X translocations in *S. coprophila* showed that the precocious chromosome in meiosis II of spermatogenesis is the same chromosome with is later eliminated from the soma and germ line of the embryo (Crouse 1943); this could also be deduced for *S. reynoldsi* where the four ordinary chromosomes are morphologically distinguishable (Crouse 1943). Only crosses of gynogenic mothers (X'X) with males carrying an X translocation yield viable progeny; in crosses of X translocation males with androgenic mothers (XX), both copies of the precocious chromosome (e.g., II^X in stock T1) are eliminated from the embryonic soma resulting in an unbalanced chromosome constitution that is lethal (Crouse 1943, 1960 b). Moreover, the X translocation studies indicated that a balanced chromosome set is not prerequisite for normal early development, as the zygote can be hypoploid or aneuploid, and it is sufficient to correct this condition to the euploid XX or XO somatic complement by embryonic somatic chromosome elimination (Crouse 1960 b). For example, when translocation T1 males are crossed with X'X females, the zygote is aneuploid because it has a duplication for part of chromosome II (since II^X was the precocious chromosome in spermatogenesis), and it has a deficiency of the X (the zygote has only two and not three sex chromosomes because they were not precocious in spermatogenesis); this is corrected in the female embryo which loses one copy of II^X by somatic elimination and returns to a balanced chromosome set (Crouse 1960 b).

When 3:1 disjunction occurs in eggs of females carrying an X translocation ($X'X^T$ or XX^T eggs), then the egg receives either two sex chromosomes or no sex chromosomes instead of the usual case of one sex chromosome (Crouse 1960 a). In one example, when these eggs are fertilized by wild type sperm, if the zygote had already been determined to be a female by a gynogenic mother, then only one paternal X chromosome is eliminated (even though the sperm donated two identical paternal X chromosomes, as is usual for *Sciara*) (Crouse 1960 a). Therefore, a nullo-X egg from a $X'X^T$ mother will result in an XO chromosome constitution after this elimination; this embryo is patroclinous for the single X it retains, since the X was obtained from its father (Crouse 1960 a). On the other hand, in a second example, when these eggs are fertilized by wild-type sperm, if the zygote had already been determined to be a male by an androgenic mother, then two paternal X chromosomes will be eliminated. Therefore, a double X egg from an XX^T mother will result in an XX chromosome constitution after this elimination; this embryo is matroclinous for the two X's it retains, since both X's came from its mother (Crouse 1960 a). In the first example a gynogenic mother gives rise to unexpected sons ("exceptional" male offspring that are patroclinous), and in the second example an androgenic mother gives rise to unexpected daughters ("exceptional" female offspring that are matroclinous). This study of exceptional offspring demonstrated that the embryonic soma does not count the total number of X's in the cell, but rather just counts the number of paternal X's it was predestined to eliminate (Crouse 1960 a).

5 X Chromosome Elimination from the Germ Line of the Embryo

By 7–8 h after egg-laying, the blastoderm stage is reached: all nuclei are at the periphery of the embryonic syncytium and cell membranes begin to form (DuBois 1932 a, 1933). Granules that are important for the formation of germ cells in other animals have not been seen in the posterior portion of *Sciara* (DuBois 1933). After 8–9 divisions in the embryo (at which time the somatic chromosome eliminations have been completed), the germ cells migrate from the posterior poleplasm to the site of the gonad (Berry 1939, 1941, Rieffel and Crouse 1966). The germ cells remain in interphase during this migration (Berry 1939); this is part of a 6-day resting stage for the germ cells, which do not divide from 48 h after

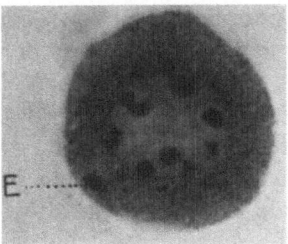

Fig. 12. Photograph of a germ cell of *Sciara ocellaris* showing eight chromosomes in the nucleus and one chromosome (**E**) expelled from the nucleus and now seen in the cytoplasm. (Berry 1941)

oviposition until the beginning of the second larval instar (Berry 1941, Rieffel and Crouse 1966).

A most remarkable chromosome elimination occurs at 33–34 h at 25 °C after oviposition in *S. ocellaris* (Berry 1941), or 65–72 h at 20 °C after oviposition in *S. coprophila* that is at least 48 h since the last mitotic division of the approximately 30 germ cells (Rieffel and Crouse 1966). This event was first noted by Berry (1939) for *S. ocellaris*, which lacks L chromosomes; he observed one of the three sex chromosomes to "pass through" (!) the nuclear membrane into the cytoplasm where it remained for about three days and then degenerated. This event occurs in the germ line of both sexes to reduce the zygotic complement of three sex chromosomes down to the germ line complement of two sex chromosomes; it is a synchronous process for all germ cells (Berry 1939). Budding of the nuclear membrane was not seen (Berry 1939); the nuclear membrane of the interphase germ cell is presumably intact, and all of a sudden one sees an X chromosome in the cytoplasm, whereas previously it had been in the nucleus (Fig. 12). Actually the germ line nuclei are not in true interphase during this period, because all the chromosomes are still condensed and can be individually counted (Berry 1941). Since it is known genetically that both a maternally and a paternally derived sex chromosome remain in the germ line, it must be the second copy of the paternal X that is eliminated from the germ cells. Is selection random for which paternal X is eliminated, and if not, why not (Berry 1941)? How does the germ cell identify only one of the seemingly identical two paternal X chromosomes to be eliminated (Berry 1941)? *Sciara* presents the only known case of germ line and not just somatic elimination of chromosomal material (Berry 1941).

The paternal X that is eliminated from the germ line had been a member of the precocious dyad in metaphase II of spermatogenesis. Using reciprocal X translocations, Crouse (1965) showed that an error can occur whereby both sister chromatids instead of just one chromatid of the precocious dyad of the spermatocyte will be eliminated from the germ line of the embryo. She used two stocks, T7 and T26, which both have a reciprocal translocation between the X and chromosome IV. Androgenic females heterozygous for the translocation (X X^T) produce a few aneuploid eggs, presumably due to adjacent instead of alternate nondisjunction (or else from alternate nondisjunction coupled with a cross-over between the swollen marker on the X and the point of translocation). These aneuploid eggs were fertilized by sperm from translocation males, to result in male progeny with a balanced soma (X_{mat}^{IV}, IV_{mat}, IV_{pat}^{X}) after embryonic elimina-tion of the two copies of the precocious chromosome from the father (X^{IV}). How-ever, the germ line should be triploid for one end of chromosome IV and haploid for one end of the X, since it normally keeps one copy of the precocious chromo-some from the father (X^{IV}). Crouse (1965) observed two kinds of primary sperma-tocytes in each testis of these male progeny: some had the usual eight ordinary chromosomes plus some L chromosomes, but others had only seven ordinary chromosomes plus some L chromosomes. The latter group were deduced to de-rive from an aberrant event in which both instead of just one copy of the preco-cious chromosome (X_{pat}^{IV}) was eliminated from the embryonic germ line (Crouse 1965). She was able to confirm that the missing eighth chromosome had been paternal in origin because she found examples of primary spermatocyte buds with only three instead of four paternal chromosomes (Crouse 1965).

During (*S. coprophila*) or just after (*S. ocellaris*) chromosome elimination from the germ line, differential staining has been observed between the two sets of ordinary chromosomes: the paternal homologs are light-staining, while the ma-ternal homologs are condensed (Rieffel and Crouse 1966). Could this be a time when the imprint of homolog origin is reversed? The maternal homolog set re-tained by the primary spermatocyte must be changed to a paternal imprint prior to fertilization (Rieffel and Crouse 1966). Then, in a male embryo the paternal imprint of these chromosomes can be retained, but in a female embryo the pater-nal imprint of these chromosomes can be reversed any time after embryonic chro-mosome elimination (Rieffel and Crouse 1966).

6 Germ Line Limited "L" Chromosomes

6.1 L Chromosome Elimination from the Germ Line

All the L chromosomes are eliminated from the somatic lineage, by remaining at the center of the anaphase spindle in the fifth (sixth) division of the embryo (Sect. 4.2). By contrast, only some of the L chromosomes are eliminated from the germ line. The egg generally has one L chromosome after meiosis, but the sperm carries 0–4 L chromosomes (78% have 2 L chromosomes) (Rieffel and Crouse 1966). Elimination of some of the L chromosomes must occur in the germ line to prevent their increasing accumulation with each generation (Metz 1938). Therefore, 0–3 L chromosomes are eliminated, to reduce the germ line comple-ment to 2 L chromosomes (Rieffel and Crouse 1966). L chromosomes are elim-inated at the same times as one paternal X from the germ line (i.e., 65–72 h at 20 °C after oviposition in *S. coprophila*); the L chromosomes may either precede or follow the X to enter the cytoplasm, presumably by passage "through" the nu-clear membrane (Rieffel and Crouse 1966). Although oocytes generally have 2 L chromosomes (and therefore the haploid egg has 1 L chromosome), the number of L chromosomes in spermatocytes is variable (Crouse et al. 1971). The latter may be explained by mitotic nondisjunction during the six gonial divisions, which occur after elimination of all but 2 L chromosomes from the embryonic germ line and before spermatogenesis (Crouse et al. 1971).

6.2 L Chromosomes and the Sex Ratio

Normally *S. impatiens* is monogenic and has L chromosomes, but a strain arose in the laboratory which became digenic, and this was correlated with the appearance of L centric fragments and ultimately no L chromosomes (Crouse et al. 1971). This nullo-L strain seems to have arisen due to a mistake in germ line elimination of L chromosomes (Crouse et al. 1971). After half a year, the ratio of bisexual progeny shifted in the nullo-L strain, and only sons were produced, so the nullo-L strain could no longer be perpetuated (Crouse et al. 1971). These observations suggested that somehow L chromosomes seem to affect the sex ratio (Crouse et al. 1971). However, inspection of Table 2 shows that the presence or absence of L chromosomes cannot be correlated with whether a species of *Sciara* will be monogenic or digenic (Metz 1929 b, 1938). All species which are only monogenic have an L chromosome, but there are species that are only digenic with L chromosomes and other digenic species that lack L chromosomes. Are some of the genes on the L chromosomes transferred to the regular chromosomes in those species that lack L chromosomes (Rieffel and Crouse 1966)? Finally, *S. simulans* can be either monogenic or digenic and both types have L chromosomes, while *S. ocellaris* can be either monogenic or digenic and both types lack L chromosomes (Metz 1938). Therefore, why the sex determination pattern changed in *S. impatiens* from monogenic to digenic when it lost the L chromosomes remains obscure.

The occurrence of the nullo-L strain of *S. impatiens* allowed experiments to demonstrate that the L chromosomes are not imprinted (Crouse et al. 1971). When the plus L strain was crossed with the minus L strain, in all cases the L chro-

Table 2. L Chromosomes and Sex Ratio in *Sciara*

	L Chromo-somes	Comments	Reference
Monogenic species			
S. coprophila Lintner	+		Metz (1938)
S. impatiens Johannsen	+		Metz (1938)
S. varians Johannsen	+		Metz (1938)
S. subtrivialis Petty	+		Metz (1938)
Digenic species			
S. pauciseta Felt	+		Metz (1938)
S. prolifica Felt	+		Metz (1938)
S. coprophila Lintner		Mutant strain	Metz and Moses (1928), Metz (1929b, 1931a)
	One L		Reynolds (1938)
S. impatiens Johannsen		Mutant strain	Metz and Moses (1928), Metz
	−		(1931a), Crouse et al. (1971)
S. reynoldsi	−		Metz and Lawrence (1938), Metz (1938)
Monogenic and digenic species			
S. simulans Johannsen	+		Metz (1938)
S. ocellaris Comstock	−		Metz and Lawrence (1938), Metz (1938)

mosomes were not discarded in the bud of the primary spermatocyte, but were retained in the group of chromosomes at the monopole, regardless of maternal or paternal origin of the L chromosomes (Crouse et al. 1971). Furthermore, when a minus L male was crossed with a plus L female, germ line development proceeded normally even though only maternally derived L chromosomes were present (Crouse et al. 1971).

6.3 L Chromosome Condensation Cycle

It has been postulated that L chromosomes escape being imprinted because they are inactive at a time when the ordinary chromosomes are active (Rieffel and Crouse 1966). It was recognized early on that the L chromosomes differ in their condensation cycle from the ordinary chromosomes in the germ line (e.g., the L chromosomes are condensed while the ordinary chromosomes are light-staining in early prophase I of spermatogenesis) (Metz 1926c). Additional observations on differential condensation of L versus ordinary chromosomes were made subsequently for other developmental stages; the L chromosomes become diffuse in the germ line at 96 h after oviposition, just shortly after germ line chromosome elimination (Rieffel and Crouse 1966). The L chromosomes also decondense during interphase between the two meiotic divisions of spermatogenesis (Amabis et al. 1979). ^{3}H-uridine uptake studies are needed to prove whether L chromosomes are transcriptionally active at these times. If the L chromosomes do code for RNA, what is the function of this RNA, and how can some species of *Sciara* survive perfectly well without any L chromosomes (Table 2)? Is L chromosome activity related to the imprinting mechanism, as the times of L chromosome diffuseness correlate with when the imprint may be reversed (see the end of Sect. 5)?

^{3}H-thymidine uptake showed that the L chromosomes seem to replicate their DNA later in the cell cycle than the ordinary chromosomes, as is generally characteristic for heterochromatin (Rieffel and Crouse 1966, Amabis et al. 1979).

7 Sex Determination

7.1 Exceptional Offspring Among Unisexual Progeny

Monogenic strains of *Sciara* produce unisexual progeny, but occasionally an unexpected "exceptional" offspring of the other sex will appear. The exceptional females that appear among their male siblings are always XX (male producers) and never X'X (female producers) (Metz and Moses 1928, Metz 1929b). Using the recessive marker swollen on the X, some exceptional males that were X'O in their soma were found among female siblings; usually these exceptional X'O males are sterile, but a few are fertile (Metz and Schmuck 1929b). Since the exceptional X'O male looks the same morphologically as the regular XO male, the X' chromosome must be very similar to the X chromosome (Metz and Schmuck 1929b). Using sex-linked recessive markers in a unisexual line of *S. ocellaris*.

Davidheiser (1943) concluded that exceptional males that inherit a maternal X are fertile, whereas exceptional males with a paternal X lack sperm and are sterile. By coupling genetic markers with X translocations, Crouse (1960a) was able to show that exceptional offspring are the result of nondisjunction in the egg; she found an increased percentage in exceptional offspring when chromosome rearrangements decreased the amount of sex chromosome synapsis in the egg. If she made the sex chromosome translocation homozygous (X^T X^T) to restore full synapsis, then no exceptional progeny were seen (Crouse 1960a). It is clear from her studies that the X' chromosome dictates the elimination of only one X from the soma and germ line, regardless of the total number of X chromosomes present (Crouse 1960a). For example, a nullo-X egg arising from nondisjunction in a X'X^T mother will already have been conditioned by the X' chromosome, and after fertilization one of the two identical paternal X chromosomes will be eliminated from both the soma and germ line. The result will be an XO chromosome constitution, and this exceptional male will be patroclinous for his paternally derived X, and probably will be sterile as his germ line is only XO instead of the usual male germ line constitution of XX (Crouse 1960a). In another example, a double X egg arising from nondisjunction in a XX^T mother will already be conditioned to lose two sex chromosomes in the embryo, since no X' chromosome was present in this mother. Thus, upon fertilization, both identical paternal X chromosomes will be eliminated in the soma and germ line, resulting in an XX chromosome constitution; this exceptional female will be fertile because she has the usual condition of two sex chromosomes in her germ line.

7.2 Bisexual Strains Derived from Monogenic Species

The occurrence of exceptional males and females shows that the egg has the potential to develop into either sex (Metz 1938). The sex ratio is rarely 1:1 in bisexual strains of *Sciara*, and a bisexual strain could be considered to be a monogenic strain with a greatly increased number of exceptional progeny (Metz 1931a, 1938). This would predict a common sex determination mechanism for both monogenic and digenic strains of *Sciara* (Metz 1938). *S. coprophila* is normally monogenic, but an exceptional female gave rise to a digenic strain in the laboratory (Metz and Moses 1928, Metz 1931a). Since the exceptional female was XX, she should have been androgenic, but instead progeny of both sexes were obtained. Presumably nondisjunction had occurred in the exceptional XX female to yield a double X egg; when the XX egg was fertilized by a standard sperm, the two identical paternal X chromosomes were eliminated in the soma, reducing it to XX (a female soma, matroclinous for both X chromosomes) (Reynolds 1938). However, only one paternal X would be eliminated from the germ line, resulting in triple X oocytes, which after meiosis gave XX eggs that would develop into females after fertilization, and single X eggs that would develop into males after fertilization (Reynolds 1938). The female offspring would be like their mothers and the process would repeat itself each generation a bisexual daughter was mated. However, a male from the bisexual line when mated with a unisexual female resulted in unisexual progeny, confirming that it is the three X chromosomes on the bisexual female germ line that were responsible for bisexual offspring (Reynolds 1938). Notice that no X' chromosome is present in this bisexual strain, and the standard chromosome eliminations of two X chromosomes lost from the soma and one X chromosome lost from the germ line occur as predicted for embryos from a mother that lacked an X' chromosome. In this case, the identical sex chromosome eliminations occur in both male and female embryos, and the sex is determined by whether the egg was single or double X.

Reynolds (1938) performed crosses of bisexual females from this mutant strain of *S. coprophila* with males from the standard monogenic line of *S. coprophila*. Each autosome of the male carried a marker, and Reynolds (1938) found that each autosome from the digenic parent could be replaced by a marked autosome from the monogenic parent, but bisexual reproduction was maintained. The conclusion is that the autosomes do not control the sex ratio of the progeny (Reynolds 1938).

In a cross of *S. ocellaris* with *S. reynoldsi*, a change in shape of one of the autosomes was seen, and it was suggestive but unresolved if this was related to bisexual versus unisexual progeny (Crouse 1939). The data of Reynolds (1938) discussed above would be counter to any relationship of autosome shape with the sex ratio of the progeny.

7.3 Sex Determination of the Gonad

Since the XX chromosome constitution occurs in the germ lines of both males (with XO soma) and females (with XX or X'X soma), it follows that gonad differentiation cannot be determined directly by the germ line chromosomes; it is more likely that the soma which differs between males (XO) and females (XX or X'X) will govern differentiation of the gonad (DuBois 1933, Metz 1938, Crouse 1960a). However, a study of gynandromorphs was contrary to this deduction: aberrant chromosome elimination events may cause somatic gynandromorphs but the gonads were either both testes or both ovaries, suggesting that gonad sex is not determined by the soma (DuBois 1932a, 1933). On the other hand, gynandromorphs that result from a cross of a *S. ocellaris* female with a *S. reynoldsi* male do have mixed gonads of one testis and one ovary per animal, supporting the view that the somatic chromosome constitution does indeed influence germ line differentiation (Metz and Lawrence 1938, Lawrence and Crouse referred to in Metz 1938 and in Crouse 1943).

Usually the males of *Sciara* are matroclinous for their sex-linked genes: both paternal X chromosomes are eliminated from the embryonic soma and only the single maternal X chromosome is retained. Patroclinous XO males can be obtained, but are routinely sterile; the paternal X is eliminated at meiosis I of spermatogenesis and no maternal X is present to act as the precocious chromosome in meiosis II (Crouse 1943, 1960a). Using reciprocal X translocations (T7 and T26, referred to in Sect. 5), it was possible to construct males that were euploid in their soma but aneuploid in their germ line; these males were viable and fertile (Crouse 1965). In these males the germ line is triploid for one end of chromosome IV and haploid for one end of the X; moreover, the segment of the X distal to the translocation is patroclinous instead of matroclinous. This semi-patroclinous male has a maternal translocation chromosome that acts as the precocious chromosome in meiosis II and active sperm are subsequently formed (Crouse 1965). Although the male germ line can tolerate aneuploidy, apparently the male soma cannot and it is lethal (Crouse 1965). On the other hand, both the soma and germ line of females can tolerate an aneuploid condition and normal ovary development proceeds (Crouse 1965).

Further genetic studies in *Sciara* would be useful to identify genes that influence gonad differentiation, and also genes that govern the unique mechanism of sex determination. It would be interesting to know if any of these genes in *Sciara* have counterparts in the sex determination genes of *Drosophila* (Baker and Belote 1983), which is a more conventional organism.

Acknowledgments. The author's laboratory is registered with the Genetics Society of America as housing the historical collection of research notebooks and microscope slides of Dr. Charles W. Metz (deceased), and also is registered as the stock-keeping center for all extant lines of *Sciara coprophila.* The author is grateful to the family of Dr. C. W. Metz for the historical collection, and is grateful to Dr. Helen V. Crouse for the X translocation and mutant lines of *S. coprophila,* and her instruction on their care. Most of what we know today about the cytogenetics of *Sciara* was uncovered by the beautiful experiments performed by Dr. H. V. Crouse and also by Dr. C. W. Metz. Thanks are due to Carol King and Jean Waage for assistance in typing this review.

References

Abbott AG, Gerbi SA (1981) Spermatogenesis in *Sciara coprophila.* II. Precocious chromosome orientation in meiosis II. Chromosoma 83:19–27

Abbott AG, Hess JE, Gerbi SA (1981) Spermatogenesis in *Sciara coprophila.* I. Chromosome orientation on the monopolar spindle of meiosis I. Chromosoma 83:1–18

Amabis JM, Reinach FC, Andrews N (1979) Spermatogenesis in *Trichosia pubescens* (Diptera: Sciaridae). J Cell Sci 36:199–213

Baker BS, Belote JM (1983) Sex determination and dosage compensation in *Drosophila melanogaster.* Annu Rev Genet 17:345–393

Basile R (1970) Spermatogenesis in *Rhynchosciara angelae* Nonato and Pavan, 1951. Rev Bras Biol 30:29–38

Berry RO (1939) Observations on chromosome elimination in the germ cells of *Sciara ocellaris.* Proc Natl Acad Sci USA 25:125–127

Berry RO (1941) Chromosome behavior in the germ cells and development of the gonads in *Sciara ocellaris.* J Morphol 68:547–583

Bird AP (1984) DNA methylation – how important in gene control? Nature (London) 307:503–504

Boveri TH (1904) Ergebnisse über die Konstitution der chromatischen Substanz des Zellkerns. Fisher, Jena

Brown SW, Chandra HS (1977) Chromosome imprinting and the differential regulation of homologous chromosomes. In: Goldstein L, Prescott DM (eds) Cell biology: A comprehensive treatise, vol I. Academic Press, London New York, pp 109–189

Carson HL (1946) The selective elimination of inversion dicentric chromatids during meiosis in the eggs of *Sciara impatiens.* Genetics 31:95–113

Chandra HS, Brown SW (1975) Chromosome imprinting and the mammalian X chromosome. Nature (London) 253:165–168

Crouse HV (1939) An evolutionary change in chromosome shape in *Sciara.* Am Nat 73:476–480

Crouse HV (1943) Translocations in *Sciara*; their bearing on chromosome behavior and sex determination. Univ Mo Res Bull 379:1–75

Crouse HV (1960a) The nature of the influence of X-translocations on sex of progeny in *Sciara coprophila*. Chromosoma 11:146–166

Crouse HV (1960b) The controlling element in sex chromosome behavior in *Sciara*. Genetics 45:1429–1443

Crouse HV (1961) X-ray induced sex-linked recessive lethals and visibles in *Sciara coprophila*. Am Nat 95:21–26

Crouse HV (1965) Experimental alterations in the chromosome constitution of *Sciara*. Chromosoma 16:391–410

Crouse HV (1977) X heterochromatin subdivision and cytogenetic analysis in *Sciara coprophila* (Diptera, Sciaridae) I. Centromere localization. Chromosoma 63:39–55

Crouse HV (1979) X heterochromatin subdivision and cytogenetic analysis in *Sciara coprophila* (Diptera, Sciaridae). Chromosoma 74:219–239

Crouse HV, Smith-Stocking H (1938) New mutants in *Sciara* and their genetic behavior. Genetics 23:275–282

Crouse HV, Brown A, Mumford BC (1971) L-chromosome inheritance and the problem of chromosome "imprinting" in *Sciara* (Sciaridae, Diptera). Chromosoma 34:324–339

Crouse HV, Gerbi SA, Liang CM, Magnus L, Mercer IM (1977) Localization of ribosomal DNA within the proximal X heterochromatin of *Sciara coprophila* (Diptera, Sciaridae). Chromosoma 64:305–318

Davidheiser B (1943) Inheritance of the X chromosome in exceptional males of *Sciara ocellaris* (Diptera). Genetics 28:193–199

Doyle WL (1933) Observations on spermiogenesis in *Sciara coprophila*. J Morphol 54:477–491

DuBois AM (1932a) A contribution to the morphology of *Sciara*. J Morphol 54:161–195

DuBois AM (1932b) Elimination of chromosomes during cleavage in the eggs of *Sciara* (Diptera). Proc Natl Acad Sci USA 18:352–356

DuBois AM (1933) Chromosome behavior during cleavage in the eggs of *Sciara coprophila* (Diptera) in the relation to the problem of sex determination. Z Wiss Biol Abt B–Z Zellforsch Mikrosk Anat 19:595–614

Eastman EM, Goodman RM, Erlanger BF, Miller OJ (1980) 5-Methylcytosine in the DNA of the polytene chromosomes of the Diptera *Sciara coprophila*, *Drosophila melanogaster* and *D. persimilis*. Chromosoma 79:225–239

Fahmy O (1949) A new type of meiosis in *Plastosciara pectiventris* (Nematocera, Diptera) and its evolutionary significance. Proc Egypt Acad Sci 5:12–42

Himes M, Crouse HV (1961) The contribution of the limited chromosome to the DNA of the giant nurse cells of *Sciara*. Abstr Annu Meet Am Soc Cell Biol, Chicago IL

Johannsen OA (1909) The fungus gnats of North America part I. Maine Agric Exp Stn Bull 172:209–276

Johannsen OA (1912) Mycetophilidae of North America. Maine Agric Exp Stn Bull 200:57–146

Kubai DF (1982) Meiosis in *Sciara coprophila*: Structure of the spindle and chromosome behavior during the first meiotic division. J Cell Biol 93:655–669

Makielski SK (1966) The structure and maturation of the spermatozoa of *Sciara coprophila*. J Morphol 118:11–42

Metz CW (1925a) Chromosome behavior in *Sciara* (Diptera). Anat Rec 31:346–347

Metz (1925b) Chromosomes and sex in *Sciara*. Science 61:212–214

Metz CW (1926a) An apparent case of monocentric mitosis in *Sciara* (Diptera). Science 63:190–191

Metz CW (1926b) Genetic evidence of a selective segregation of chromosomes in *Sciara* (Diptera). Proc Natl Acad Sci USA 12:690–692

Metz CW (1926c) Chromosomal studies on *Sciara* (Diptera). I. Differences between the chromosomes of the two sexes. Am Nat 60:42–56

Metz CW (1927) Chromosome behavior and genetic behavior in *Sciara* (Diptera). II. Genetic evidence of selective segregation in *S. coprophila*. Z Indukt Abstamm Vererbungsl 45:184–200

Metz CW (1928) Genetic evidence of a selective segregation of chromosomes in a second species of *Sciara* (Diptera). Proc Natl Acad Sci USA 14:140–141

Metz CW (1929a) Evidence that unisexual progenies in *Sciara* are due to selective elimination of gametes (sperms). Am Nat 63:214–228

Metz CW (1929b) Sex determination in *Sciara*. Am Nat 63:487–496

Metz CW (1929c). Selective segregation of chromosomes in males of a third species of *Sciara*. Proc Natl Acad Sci USA 15:339–343

Metz CW (1930) A possible alternative to the hypothesis of selective fertilization in *Sciara*. Am Nat 64:380–382

Metz CW (1931a) Unisexual progenies and sex determination in *Sciara*. Q Rev Biol 6:306–312

Metz CW (1931b) Chromosomal differences between germ cells and soma in *Sciara*. Biol Zentralbl 51:119–124

Metz CW (1933) Monocentric mitosis with segregation of chromosomes in *Sciara* and its bearing on the mechanism of mitosis. I. The normal monocentric mitosis. II. Experimental modification of the monocentric mitosis. Biol Bull 64:333–347

Metz CW (1934) Evidence that in *Sciara* the sperm regularly transmits two sister sex chromosomes. Proc Natl Acad Sci USA 20:31–36

Metz CW (1936) Factors influencing chromosome movements in mitosis. Cytologia 7:219–231

Metz CW (1938) Chromosome behavior, inheritance and sex determination in *Sciara*. Am Nat 72:485–520

Metz CW (1957) Interactions between chromosomes and cytoplasm during early embryonic development in *Sciara* (Diptera). Biol Bull 113:323

Metz CW, Lawrence EG (1938) Preliminary observations on *Sciara* hybrids. J Hered 29:179–186

Metz CW, Moses MS (1926) Sex determination in *Sciara* (Diptera). Anat Rec 34:170

Metz CW, Moses MS (1928) Observations on sex-ratio determination in *Sciara* (Diptera). Proc Natl Acad Sci USA 14:930–932

Metz CW, Schmuck ML (1929a) Unisexual progenies and the sex chromosome mechanism in *Sciara*. Proc Natl Acad Sci USA 15:863–866

Metz CW, Schmuck ML (1929b) Further studies on the chromosomal mechanism responsible for unusual progenies in *Sciara*. Tests of "exceptional" males. Proc Natl Acad Sci USA 15:867–870

Metz CW, Schmuck ML (1931a) Studies on sex determination and the sex chromosome mechanism in *Sciara*. Genetics 16:225–253

Metz CW, Schmuck ML (1931b) Differences between chromosome groups of soma and germline in *Sciara*. Proc Natl Acad Sci USA 17:272–275

Metz CW, Schmuck Armstrong L (1961) Observations on deficient chromosome groups in developing *Sciara* larvae. Growth 25:89–106

Metz CW, Smith HB (1931) Further observation on the nature of the x-prime (X') chromosome in *Sciara*. Proc Natl Acad Sci USA 17:195–198

Metz CW, Ullian SS (1929) Genetic identification of the sex chromosomes in *Sciara* (Diptera). Proc Natl Acad Sci USA 15:82–85

Metz CW, Moses MS, Hoppe EN (1926) Chromosome behavior and genetic behavior in *Sciara* (Diptera). I. Chromosome behavior in the spermatocyte divisions. Z Indukt Abstamm Vererbungsl 42:237–270

Moses MS, Metz CW (1928) Evidence that the female is responsible for the sex ratio in *Sciara* (Diptera). Proc Natl Acad Sci USA 14:928–930

Phillips DM (1966a) Observations on spermiogenesis in the fungus gnat *Sciara coprophila*. J Cell Biol 30:477–497

Phillips DM (1966b) Fine structure of *Sciara coprophila* sperm. J Cell Biol 30:499–517

Phillips DM (1967) Giant centriole formation in *Sciara*. J Cell Biol 33:73–92

Phillips DM (1970) Insect sperm: their structure and morphogenesis. J Cell Biol 44:243–277

Retzius G (1904) Zur Kenntnis der Spermien der Evertebraten. Biol Untersuch 11:1–32

Reynolds JT (1938) Sex determination in a "bisexual" strain of *Sciara coprophila*. Genetics 23:203–220

Rieffel SM, Crouse HV (1966) The elimination and differentiation of chromosomes in the germ line of *Sciara*. Chromosoma 19:231–276

Sager R, Kitchen R (1975) Selective silencing of eukaryotic DNA. Science 189:426–433

Schmuck ML (1934) The male somatic chromosome group in *Sciara pauciseta*. Biol Bull 66:224–227

Schmuck ML, Metz CW (1931) A method for the study of chromosomes in entire insect eggs. Science 74:600–601

Schmuck ML, Metz CW (1932) The maturation divisions and fertilization in eggs of *Sciara coprophila*: Lintner. Proc Natl Acad Sci USA 18:349–352

Smith-Stocking H (1936) Genetic studies on selective segregation of chromosomes in *Sciara coprophila* Lintner. Genetics 21:421–443

Steffan WA (1966) A generic revision of the family Sciaridae (Diptera) of America North of Mexico. Univ Calif Publ Entomol 44:1–77

Sturtevant AM, Beadle GW (1936) The relations of inversions in the X chromosome of *Drosophila melanogaster* to crossing over and disjunction. Genetics 21:554–604

Wei L-H, Erlanger BF, Eastman EM, Miller OJ, Goodman R (1981) Inverse relationship between transcriptional activity and 5-methylcytosine content of DNA in polytene chromosomes of *Sciara coprophila*. Exp Cell Res 135:411–415

White MJD (1973) Animal cytology and evolution, 3rd edn. Cambridge Univ Press, Cambridge, pp 516–523

Wilson EB (1925) The cell in development and heredity, 3rd edn. MacMillan, New York, pp 168–172

Zilz ML (1970) In vitro male meiosis in the fungus gnat (*Sciara*) with analysis of chromosome movements during anaphase I and II. PhD thesis, Wayne State Univ. Univ Microfilms, Ann Arbor, pp 1–122

Molecular Reorganization
During Nuclear Differentiation in Ciliates

G. Steinbrück [1]

1 Introduction

Differentiation denotes the process by which cells of multicellular organisms become different from one another. The closely related process by which nuclei become functionally and/or structurally different from one another can be called nuclear differentiation. Nuclear differentiation is a well-known, but often poorly understood phenomenon. For instance, large differences exist between gamete nuclei, e.g., sperm nuclei and somatic nuclei of the same organism, with regards to nuclear size, chromatin condensation, transcriptional activity, structure of the nuclear lamina, and many other characteristics. But differences in nuclear structure and in the organization of the genetic material also occur between different somatic nuclei. Differing amounts of eu- and heterochromatin, DNA rearrangements during cellular development, or amplification of certain genes may give rise to the existence of nuclei which differ from one another despite their common origin from one zygote nucleus.

It is less well-known that comparable nuclear differentiation can be found quite frequently also among unicellular organisms. In this case structurally and functionally different nuclei develop either in different stages of the life cycle of a unicellular organism or in some cases even within one cell.

The most frequent type of nuclear differentiation found in unicellular eukaryotes is a sequential differentiation of nuclei in different stages of the cellular life cycle (Grell 1973). In the alga *Acetabularia* for instance, very obvious differences in structure and function can be detected between the large primary nucleus of the vegetative cell and the numerous small secondary nuclei found in the cysts (Brachet and Bonotto 1970, Werz 1974, Wollgiehn 1982).

Among protozoans comparable differences are known in several groups. Differences in several aspects seem to exist between the large, perhaps polygenomic, nuclei of vegetative cells and the small nuclei of the swarmers in radiolarians (Grell 1973). In homokaryotic foraminiferans sequential nuclear differentiation occurs between agamont, gamont, and gamete stages of the life cycle (l.c.). In sporozoans (l.c.) even more complex patterns of sequential nuclear differentiation can be found.

Another type of nuclear differentiation – a differentiation between two different nuclei within one cell – takes place only in heterokaryotic foraminiferans and

[1] Universität Tübingen, Institut für Biologie III, Abteilung Zellbiologie, Auf der Morgenstelle 28, 7400 Tübingen, FRG.

Results and Problems in Cell Differentiation 13
Germ Line – Soma Differentiation (Ed. by W. Hennig)
© Springer-Verlag Berlin Heidelberg 1986

in ciliates. This type of nuclear differentiation results in cells which contain two types of nuclei which are structurally and functionally different. This occurrence of two different types of nuclei within one cell is often called "nuclear dualism".

Nuclear dualism in ciliates and foraminiferans has been known for a long time. In contrast, very little was known until recently about molecular differences which arise in this process, only speculations on molecular mechanisms which lead to and regulate these nuclear differentiations were in existence. But at least for ciliates this situation has changed somewhat during the last few years.

Recently, an excellent comprehensive review on "The Protozoan Nucleus" was published by Raikov (1982) which summarized the literature until 1981. Whereas Raikov's main interest centered around the ultrastructural level of nuclear morphology in Protozoan nuclei in general, I will review and discuss the most recently gained information on the molecular events occurring during nuclear differentiation in ciliates. These molecular data shall be discussed in their relation to the structural organization of the nuclei under three main aspects:

1) Which evolutionary trends can be detected in molecular reorganization during nuclear differentiation in ciliates?
2) What are the putative functions of chromosome and chromatin elimination, DNA modificiation, rearrangement, and amplification generating quite differently organized nuclei?
3) Which of the molecular reorganization processes are of general importance for the elucidation of nuclear differentiation phenomena in eukaryotes and which seem to be a peculiarity of the ciliates?

Since only very few of the about 7,000 known species of ciliates have so far been investigated by molecular methods, much of the discussion must remain highly speculative. But such a discussion might help to direct scientific interest to those groups or those open questions whose investigation might provide the greatest progress in understanding nuclear differentiation.

The whole subject will be treated as followed: After a short general description of the cytological events leading to nuclear dualism in ciliates, these events will be described in molecular detail for those groups of ciliates which have been investigated by molecular and biochemical methods in this respect or which seem especially important under this aspect. On the basis of presented experimental data, the three groups of problems cited above will be discussed.

The scanning electron micrographs shown in Fig. 1 give an impression of the general appearance of those four ciliate species which are mentioned most frequently in this article.

Fig. 1 a–d. Scanning electron micrographs of four ciliate species. **a** Ventral view of the hypotrich *Stylonychia lemnae*. **b** Ventral view of the hypotrich *Oxytricha "nova"*. The denotive characteristic of hypotrichs - the ventral and marginal cirri consisting of fused cilia – is clearly visible as well as the adoral zone of membranelles which mark the cytostome. **c** Ventral view of the hymenostomatid *Tetrahymena pyriformis* showing the cytostome and the ciliary rows. **d** Ventral view of the hymenostomatid *Paramecium tetraurelia*. The cytostome is somewhat hidden by the numerous cilia. The *bars* represent 20 µm. **a** from Ammermann and Schlegel (1983), with permission. **b, c, d** Originals kindly provided by M. Schlegel

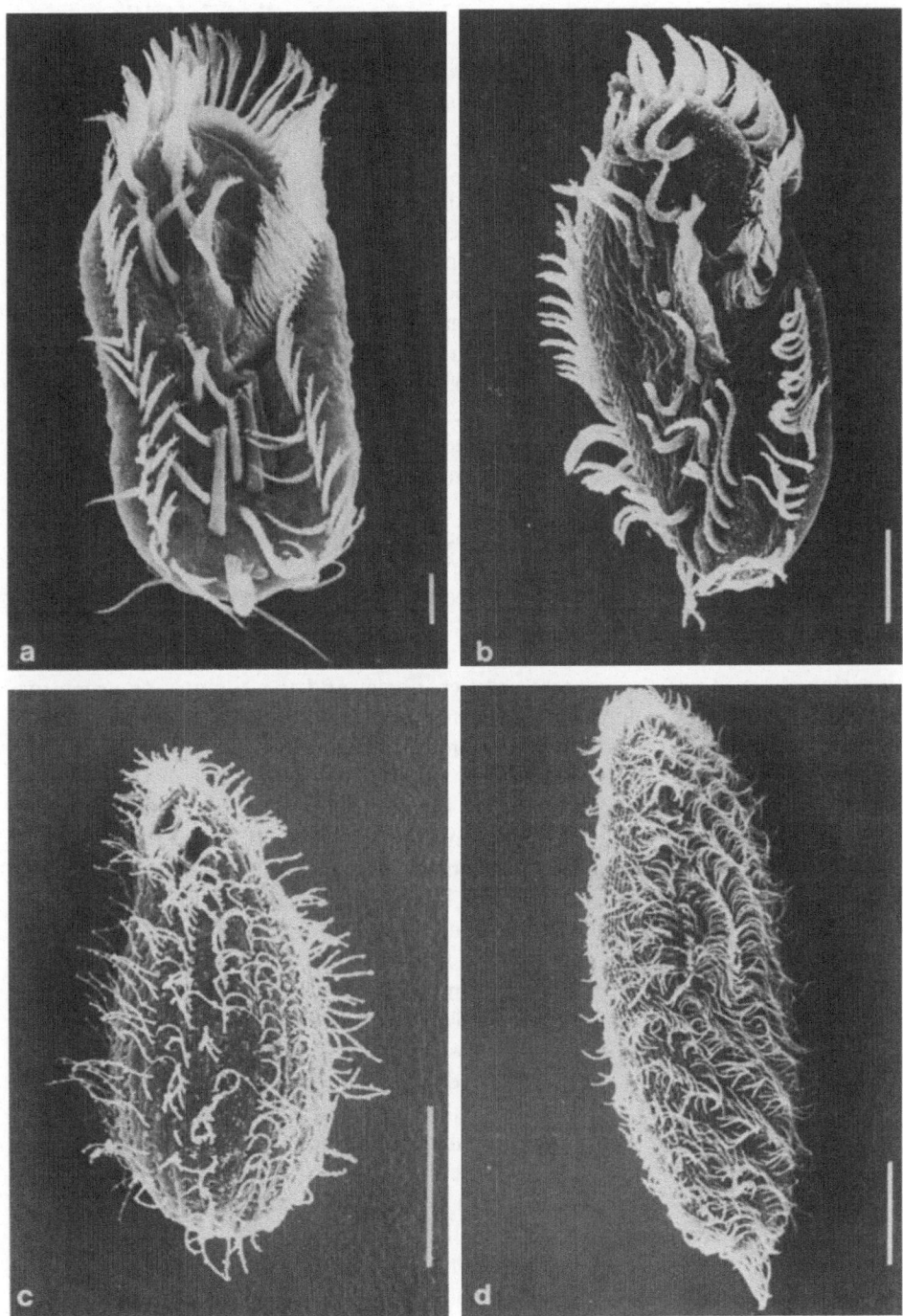

2 A General Description of Nuclear Dualism in Ciliates

Nuclear dualism is characteristic for all living ciliate species. For some time it was assumed that an exception of that rule existed in the members of the genus *Stephanopogon*. *Stephanopogon* species resemble ciliates since they move by means of "cilia" which are arranged in rows. In addition, they have a cytostome-cytopharyngeal apparatus which is used for actively feeding bacteria and small eukaryotes. Therefore, they were believed to be primitive ciliates (Raikov 1969, 1982, Corliss 1979). But their cells contain a large number of virtually identical nuclei. Consequently, *Stephanopogen* species were considered to be karyological relicts because of their homokaryotic character. They seemed to be survivors of primitive ciliates which had not reached the more advanced stage of nuclear dualism which is found in all other ciliate species (Corliss 1979). *Stephanopogon* species seemed to be "missing links" between eukaryotes with one nucleus or several identical nuclei per cell and all contemporary ciliates with their peculiar nuclear dualism. But recently it has been shown that the *Stephanopogon* species do not belong to the ciliates at all. Their "cilia" are in fact flagella and they are presumably flagellates as detailed electron microscopic investigations indicate (Lipscombe and Corliss 1982). Therefore, no exception from nuclear dualism in ciliates is known and no "missing links" between ciliates and other eukaryotic unicells have been detected so far.

The two types of nuclei occurring in every ciliate cell (rare exceptions will be mentioned later) are the generative micronuclei and the somatic or vegetative macronuclei. Macronuclei are usually large and have a considerably higher DNA content than the corresponding micronuclei. Exceptions to the rule are only found in the order Karyorelictida and in very few others. Their morphology shows a great diversity among different species. A comprehensive description of the structural variation of ciliate macronuclei is given by Raikov (1982). The assumption based on their high DNA content that macronuclei are always polyploid or polygenomic nuclei (Grell 1973, Raikov 1982) must be revised in the light of recently gained experimental evidence. This important problem will be discussed in detail later. Macronuclei have exclusively somatic functions and are very active in transcription. They contain in most cases numerous nucleoli. Macronuclear volume and DNA content seem to be positively correlated with cell size (e.g., Ammermann and Muenz 1982). The most frequent type of macronuclear division is the approximately equal division into daughter nuclei by a mechanism which is not completely understood (sometimes called amitosis, although this term does not seem adequate). But also multiple or unequal divisions occur in some species.

In addition to the macronuclei, ciliate cells contain one or several micronuclei. The micronuclei are small, mostly sperm-type nuclei containing highly condensed chromatin. They are diploid in nearly all ciliate species. The occurrence of polyploid micronuclei has been reported only for *Paramecium bursaria* (e.g., Chen 1940) and very few other species. Micronuclei contain chromosomes and are able to undergo meiosis. Virtually no transcriptional activity can be detected. The question whether they are absolutely inactive in transcription will be discussed later (see Sects. 4.1.6 and 4.3.3). The presence of micronuclei does not seem, at least in some cases, to be absolutely necessary for the vegetative life of the cells.

Micronuclei can be eliminated by irradiation or by micromanipulation. Such emicronucleated cells can survive for a long time. Micronuclei-free strains can also be found quite frequently in nature in some species. Whether in these cases some essential micronuclear functions are overtaken by the macronuclei and/or whether these cells are really absolutely free of any micronuclear sequences will be discussed later. In any case, the main function of the micronuclei seems to be a generative one. The micronuclear genomes represent the germ line genomes in ciliates. The products of their meiotic divisions are engaged in the fertilization process, called "conjugation" in ciliates.

The genetic material of the germ line genomes is transferred to the next generation during sexual reproduction. Macronuclei descend from micronuclei anew in a complex process after each conjugation. This process leading to a new macronucleus is called "macronuclear development". Figure 2 shows a schematic diagram of conjugation in ciliates.

Conjugation starts under suitable conditions with cells of differing mating types. Ciliate species may exhibit 2 to more than 100 different mating types (e.g., Ammermann 1982) depending on the species regarded. Conjugation starts with species-specific mating behavior. First, the cells touch each other and lie thereafter close together obviously connected in the anterior parts. Micronuclei in both conjugants undergo meiosis. As the result of two meiotic divisions, four nuclei arise (gone nuclei), three of which degenerate; one undergoes postmeiotic mitosis. This mitotic division of the haploid nucleus generates two pronuclei (gamete nuclei) in each cell. The two nuclei differentiate to different destinations. One of the two gamete nuclei, the so-called stationary pronucleus, remains in that cell where it originated. The other, the so-called migratory pronucleus, migrates into the partner cell. Since the migratory pronuclei of both cells behave in the same way, a mutual exchange of the two migratory pronuclei results. The next step is a fusion of the stationary nucleus with the newly invaded migratory nucleus in each cell. Thus, synkarya are formed by mutual fertilization. After this step the cells separate and are called "exconjugants" thereafter. In each exconjugant the synkaryon divides once mitotically. One of the division products becomes the new micronucleus, the other will be determined to become the new macronucleus. Meanwhile, the old macronucleus degenerates and is resorbed by the cell. After division of the synkaryon the presumptive new macronucleus is called "macronuclear anlage". It develops into the new macronucleus. Figure 3a shows a conjugating pair of cells of *Stylonychia lemnae*. Both cells are connected in this stage by a relatively broad cytoplasmic bridge. Figure 3b shows an exconjugant cell of the same species. The large macronuclear anlage which develops into the new macronucleus and two fragments of the old degenerating macronucleus are visible (cf. Fig. 2).

Thus, conjugation is a complex sequence of events which comprises meiosis, postmeiotic division, pronuclear exchange, mutual fertilization, postzygotic division(s), macronuclear determination, macronuclear differentiation, and degeneration of the old macronucleus.

These events can be varied from species to species mainly by the fact that in many species several micronuclei are present in a conjugating cell and/or that different numbers of postmeiotic divisions are included.

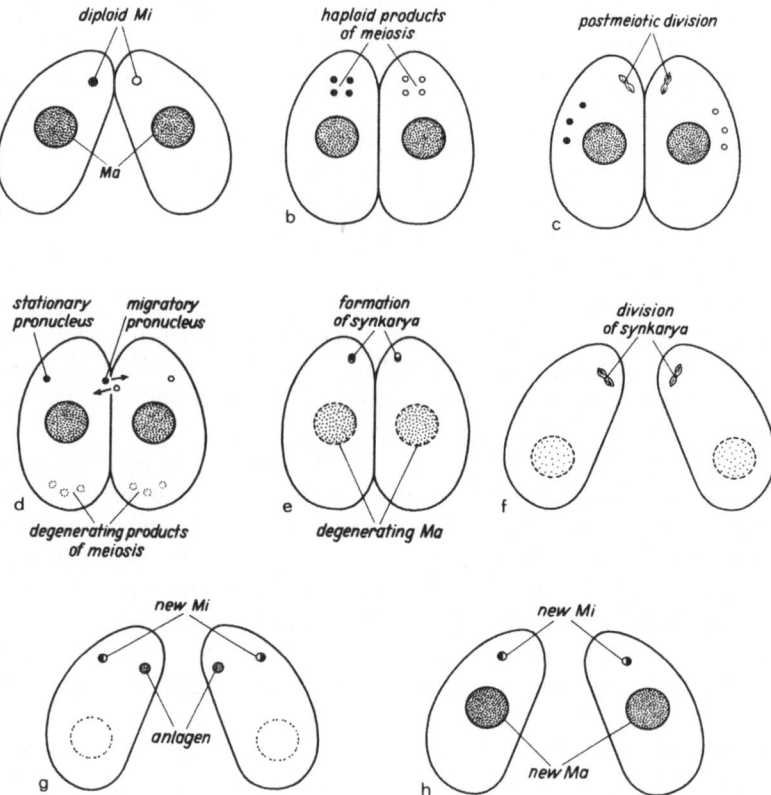

Fig. 2 a–h. Simplified diagram of conjugation in ciliates. **a** Cells of different mating types touch each other and fuse at their anterior ends. In this case, each cell contains one small diploid micronucleus (*Mi*) and one large DNA-rich macronucleus (*Ma*). **b** The conjugating cells lie side by side. Each *Mi* has undergone two successive meiotic divisions resulting in four haploid daughter nuclei. **c** One of the four haploid nuclei divides by mitosis in the so-called postmeiotic division. The other three haploid nuclei begin to degenerate. **d** The postmeiotic division results in two haploid gamete nuclei in each cell. One, the stationary pronucleus, remains in the cell where it originated. The other one, the migratory pronucleus, migrates into the partner cell. Since this occurs in both cells, a mutual exchange of migratory pronuclei results. **e** In both cells the stationary nucleus and the migratory nucleus fuse with each other (karyogamy), thus forming diploid synkarya. The old macronucleus begins to degenerate. **f** Thereafter the partner cells separate from each other and are now called exconjugants. The synkaryon of each exconjugant divides mitotically. While one of the two division products becomes the new micronucleus, the other one develops into the new macronucleus. During this development this nucleus is called the macronucleus anlage. **g** The macronucleus anlage grows and its DNA content increases by several rounds of replication in most ciliate species. The old *Ma* disintegrates more and more and at last becomes absorbed by the cytoplasm. **h** Finally, the growing *Ma* anlage has reached the DNA content and the appearance of the *Ma*. A new *Ma* has been formed

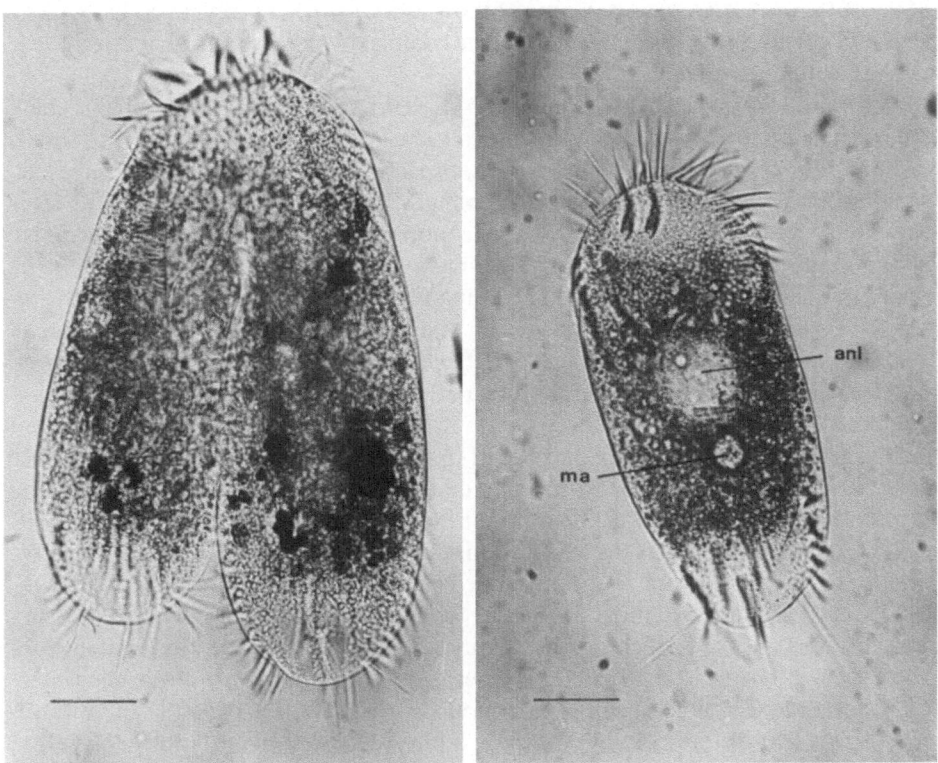

Fig. 3. a Conjugating pair of *Stylonychia lemnae* cells. Both cells are connected by a cytoplasmic bridge. **b** Exconjugant cell of *S. lemnae*. Within the cell the large macronucleus anlage (*anl*) and one smaller fragment of the old degenerating macronucleus (*ma*) are visible. The *bars* represent 30 μm. Courtesy of D. Ammermann

3 Macronuclear Development in "Primitive" Ciliates of the Order Karyorelictida

The order Karyorelictida must now be assumed to comprise the most primitive ciliates, since the *Stephanopogon* species which were earlier believed to be the most primitive ones have been recognized as not being ciliates at all as explained in the previous chapter (for discussion see Corliss and Hartwig 1977; Corliss 1979; Lipscombe and Corliss 1982). "Most primitive" means that species of this order have conserved many ancient characters from their ancestors. Those characters are either now absent in all other ciliates or have been transformed by evolution to more advanced "modern" character conditions in other groups of ciliates. Despite the fact that molecular investigations of nuclear development in karyorelictid species are just at the beginning, these species seem especially promising for a general understanding of nuclear development in ciliates. So far, mainly structural changes and quantitative DNA changes occurring during macronuclear development have been investigated in some detail in a few karyorelictid species.

Most of these studies were done by Raikov and co-workers (reviewed by Raikov 1969, 1982) and by some French protozoologists (e.g., Fauré-Fremiet 1954; Dragesco 1960; Nouzarède 1976).

Nearly all of the species belonging to this order inhabit an ecological niche – the mesopsammon – which is well-known to be a refuge of so-called living fossils (Remane 1952; Ax 1966). Living fossils are species which are assumed to have survived millions of years without great changes in their morphology and in their way of life. This is quite in contrast to the more or less rapid evolutionary change of the majority of animal species. One reason for their conservatism might be found in their particular habitat. The fauna of the mesopsammon inhabits the interstitial space between sand grains on the sea floor. Obviously, very special adaptations are required to inhabit that biotope successfully and permanently. Species belonging to quite different phyla of the animal kingdom, e.g., ciliates, hydrozoans, turbellarids, gastrotrichs, nematodes, bryozoans, tardigrads, archiannelids, crustaceans, etc., have succeeded in invading this special niche. But once adapted to this environment perhaps no major evolutionary changes are required or might even be deleterious. That might be the reason why many species of different phyla which live in the mesopsammon have conserved so much primitive organization which is not found in related species living in other habitats (l.c.).

As mentioned above most of the karyorelictid ciliate species occur in the mesopsammon; only the species of the genus *Loxodes* are living in freshwater (Raikov 1959). All of the karyorelictid species show presumably primitive characters on different levels of cellular organization (Corliss and Hartwig 1977). With respect to nuclear differentiation, the main focus shall be on macronuclear structure and development only.

It has long been known that macronuclei of karyorelictid ciliates are atypical in several aspects (Bütschli 1876; Balbiani 1890, reviewed by Raikov 1969). In many species macro- and micronuclei form nuclear clusters containing a large number of nuclei. These clusters are in some species surrounded by a common capsule or by a system of cisternae (e.g., Raikov 1972; Nouzarède 1976). In macronuclei of marine karyorelictid species protein inclusions – protein spheres or crystalloids – are found (e.g., Raikov and Dragesco 1969; Kovaleva and Raikov 1973). Most surprisingly, the macronuclei of all karyorelictid species investigated never divide (Fauré-Fremiet 1954; Raikov 1976; Nouzarède 1976). This is a unique character not found in any other ciliate group. But also in karyorelictids only the macronuclei are transcriptionally active and contain visible nucleoli (Torch 1964; Ron and Urieli 1977). After cell division the characteristic number of macronuclei which are distributed randomly during cytokinesis is restored by a generation of new macronuclei from micronuclei. Thus, in contrast to all other ciliates, macronuclear development can take place in karyorelictid species after every cell division in vegetative cells. In all other ciliates macronuclear development occurs only after the sexual processes of conjugation or autogamy. Those macronuclei which are unable to divide have a very short life span. After a few cell generations they degenerate, become pycnotic, and are finally resorbed. Also this feature is in contrast to the situation found in all other ciliates where macronuclei are not renewed during the vegetative life of the cells, but only after conjugation or autogramy.

The extraordinary form of macronuclear development raises some very important questions: is the form of macronuclear development during the vegetative life cycle of karyorelictids really a primitive characteristic among ciliates compared to the form of nuclear development found after sexual processes? In other words: is the karyorelictid form of nuclear development the evolutionary precursor of nuclear development as it is found in all other ciliates?

Is the short lifetime and the inability of karyorelictid macronuclei to divide a necessary consequence of their peculiar type of development? What is the molecular basis of these obvious macronuclear deficiencies? An answer to the last question might provide the key for an understanding of the function and evolution of macronuclear development in ciliates.

No sufficient answers to these questions are currently available. But some of the experimental data described below indicate the general direction in which to investigate.

Until recently, on the basis of cytophotometric measurements, it was assumed that the macronuclei of karyorelictids are diploid and, thus, very similar in their organization to the micronuclei (Raikov 1976). This was considered to be an additional primitive character of the Karyorelictida (Corliss 1979). But careful cytophotometric measurements of macronuclear DNA content have now altered this view, at least for some species. It could be shown that the DNA content varies and deviates considerably from a diploid value. Individual macronuclei of *Loxodes magnus* have a DNA content up to 10 C (Bobyleva et al. 1980). In *Geleia orbis* the macronuclear DNA content varies from 4 C to 30 C (Nouzarède 1976). In *Trachelonema sulcata* the macronuclear DNA content accumulates in aging nuclei up to 12 C (Kovaleva and Raikov 1978). In addition, there is autoradiographic evidence for DNA synthesis in macronuclei of *Loxodes striatus* (Ron and Urieli 1977) and *Loxodes magnus* (Raikov and Morat 1977) despite their inability to divide.

It is especially interesting that a clear example of chromatin elimination during macronuclear development was detected in *Trachelonema sulcata* (Kovaleva and Raikov 1978). It must be stressed again that this macronuclear development occurs in vegetative cells as it is characteristic for the Karyorelictida. In young macronuclear anlagen of *T. sulcata* the DNA content drops to about half the value of that of the micronuclei. This decrease in DNA content is accompanied by an accumulation of fibrous-granular material in the periphery of the anlagen as shown by electron microscopy. It is assumed that the peculiar fibrogranular bodies contain chromatin which is later eliminated into the cytoplasm. Afterwards the DNA content of the anlage increases to approximately the value of the micronuclei. In old macronuclei the DNA content can rise further (see above). The increase is slow and unsynchronous and is supposed to result from differential replication or amplification of some DNA fractions.

Based on these more recent investigations of the macronuclear DNA content of karyorelictid ciliates, it seems no longer justified to assume that the Karyorelictida have diploid macronuclei. Since the DNA composition of macro- and micronuclei also seems to differ between both types of nuclei, at least in some species, termini such as "paradiploid" or "hyperdiploid" (Raikov 1982) seem inadequate and obscure the interesting phenomena.

Ribosomal DNA amplification was assumed by some authors (e.g., Kovaleva and Raikov 1978) to be the molecular mechanism of the observed increase of macronuclear DNA content. But experimental evidence that the DNA synthesis observed in the nondividing macronuclei is only due to rDNA amplification is lacking and the arguments for it (e.g., large number of nucleoli) are not convincing. Especially the large variability of the DNA content in aging macronuclei makes this assumption, as the only explanation, at least very unprobable.

But regardless of the lacking experimental evidence for the molecular mechanism, it can be concluded that during macronuclear development changes in DNA composition and DNA content occur in vegetative cells of karyorelictid ciliates. Additional variations in DNA content can be found in macronuclei during their senescence. Both changes seem to be the result of elimination and/or amplification of some nuclear DNA fractions.

The question whether these variations of DNA composition and content in karyorelictid macronuclei is a primitive character and whether it can be compared with the nuclear development after sexual processes found in other ciliates will be discussed later (Chap. 7).

4 Molecular Changes During Macronuclear Development in Hymenostomatid Ciliates

Tetrahymena and *Paramecium* species belong to the order Hymenostomatida. Some of them are among the best known and most intensively studied ciliates. Their nuclear development is better known than for any other ciliate. The most important molecular changes involved in nuclear differentiation of the hymenostomatid species are described in this section.

4.1 *Tetrahymena*

4.1.1 The Species Problem

Studies on *Tetrahymena* species have contributed important results to many fields of biology. Especially for genetic studies *Tetrahymena* offers unique advantages (for reviews see Sonneborn 1974 b; Nanney 1980). Also nuclear differentiation of *Tetrahymena* has been investigated thoroughly. However, a problem for the nonspecialist (and often for the specialist, too) exists when he wants to compare experimental data described in older literature with more recently published results. Until about 1976 the vast majority of investigations were done with the "species" *Tetrahymena pyriformis*. More recently the most frequently used organism is *T. thermophila*. It seems that the main interest has switched from one species to another. In fact, this change reflects only advances made in species identification. *T. pyriformis* was long known to be a complex of closely related, but morphologically indistinguishable, strains which do not interbreed with each other. These strains were often called varieties or syngens. According to the genetic and evolutionary definition of "species" they are true species (Mayr 1963), but they were not given Latin species names because they could be identified only by mating tests with living reference strains and not by means of classical, e.g., morphological methods of taxonomic classification (see Sonneborn 1957 for discussion). However, after the development of biochemical methods, the species of this complex were distinguishable, thus, correct Latin names could be given. The breeding strain formerly called *T. pyriformis* syngen 1 was assigned *T. thermophila* n.sp. and the most widely used amicronucleate strain, formerly called *T. pyriformis* GL, retained the name *T. pyriformis* based on the assumption that it is derived from Lwoff's (1923) original isolate (Nanney and McCoy 1976).

Furthermore, the identification of *Tetrahymena* species by electrophoretic patterns of selected enzymes (Borden et al. 1973, 1977) provided the solution of another problem. Among *Tetrahymena* a large number of asexual strains were found in laboratories as well as in collections from nature. They could not, of course, be tested in mating experiments. Isoenzyme analysis also allowed their correct identification (Borden et al. 1973). It should be mentioned that the designation of asexual strains as "species" is in conflict with the biological species definition, a problem that cannot be discussed in this context.

When biochemical methods of species identification became available, considerable confusion in laboratory strain designation could be documented (Borden et al. 1973; McCoy 1976). This was obviously caused by mislabeling or contamination of strains. All these problems must be kept in mind when comparing molecular data of nuclear development in *Tetrahymena* species. Therefore, the species names, according to Nanney and McCoy (1976), are used in this review only when the strains were clearly designated in the original publication. Only "*Tetrahymena*" is used when the strain designation is not absolutely clear or when it seems irrelevant. The term "*T. pyriformis* complex" indicates that the described facts are assumed to be valid for the whole group of species.

4.1.2 Gross Differences Between Macro- and Micronuclear Genomes

The predominantly genetic interest in *Tetrahymena* initiated and stimulated many studies on molecular organization and differences between macro- and micronuclear genomes. These studies have been reviewed recently in detail by Gorovsky (1980). In this section only the most prominent molecular differences between macro- and micronuclei shall be reviewed. In the following sections the more specialized questions of macronuclear genome organization and alterations found in specific sequences during nuclear differentiation will be treated.

Micronuclei of species of the *T. pyriformis* complex contain a diploid set of ten chromosomes. As characteristic for most ciliates the DNA content of the micronuclei is considerably smaller than that of the corresponding macronuclei. Microspectrophotometric measurements revealed a haploid DNA content of about 0.21 pg of *T. thermophila* micronuclei (Gibson and Martin 1971; Woodard et al. 1972; Seyffert 1979). In another species only half this value was found (Seyffert 1979). The macronuclear DNA content of *Tetrahymena* species was determined by several authors (Gibson and Martin 1971; Woodard et al. 1972; Doerder and DeBault 1975; Seyffert 1979). Macronuclei of *T. thermophila* contain 7.4–10.4 pg DNA corresponding to 35–50 times the C-value. Other *Tetrahymena* species have similar DNA contents.

The macronuclear DNA base composition of several *Tetrahymena* species has been determined. The $G+C$ content varies from 25% to 33% (Conner and Koroly 1973; Nanney and McCoy 1976). Thus, the variation in base composition within the *Tetrahymena* species complex is about twice the total variation observed among all the vertebrates as Nanney (1980) pointed out. Thermal denaturation profiles of macro- and micronuclear DNA have been compared (Yao and Gorovsky 1974) only in *T. thermophila*. The base compositions of both genomes deduced from the melting curves differ slightly, but not significantly. In all experiments the T_M of micronuclear DNA was lower than the T_M of the macronuclear DNA, 0.7 °C on the average. This might be an indication for either slight differences in the average base composition or for differences in the distribution of base pairs. Unfortunately, comparable data on micronuclear DNA base composition of other species are lacking.

Renaturation kinetics of both macro- and micronuclear DNA of *T. thermophila* as well as cross-hybridization experiments with both DNA's revealed that 10–20% of the micronuclear sequences are either absent or underreplicated in macronuclear genomes (Yao and Gorovsky 1974; Iwamura et al. 1979). These differences are assumed to comprise mainly moderately repeated sequences.

In buoyant density centrifugations in neutral or alkaline CsCl gradients no major differences in satellite fractions could be detected. Perhaps the micronuclear DNA has a slightly lower buoyant density than the macronuclear DNA (l.c.).

An interesting difference was found in the amounts of methylated bases. Macronuclear DNA of *T. thermophila* contains 0.6–0.8 mol% N^6-methyladenine, micronuclear DNA less than 0.1 mol%. No methylcytosine could be detected in macro- or micronuclear DNA (Gorovsky et al. 1973; Rae and Steele 1978). Whether this difference plays a role in nuclear differentiation must be tested in further experiments.

The knowledge of the size of macro- and micronuclear DNA molecules is of crucial importance for the understanding of nuclear differentiation in *Tetrahymena*. If 10–20% of the micronuclear sequences are excised and eliminated during nuclear development, one can expect to find shorter DNA molecules in macronuclei than in micronuclei. This assumption is especially justified if the eliminated sequences are distributed randomly in the micronuclear genome and are not clustered at chromosome ends or in a few loci of the genome. Evidence supporting this assumption will be presented in Sect. 4.1.5. When DNA sequences are eliminated and the generated free ends are joined subsequently by ligase action, one can expect to find molecules of similar size in both kinds of nuclei. But true rearrangements should be detectable in this case.

The size of micronuclear DNA molecules is unknown since technical problems exist in isolating sufficient quantities of undegraded micronuclear DNA free of macronuclear contamination. But it seems generally accepted that micronuclear DNA is of chromosomal size as in other eukaryotes (e.g., Kavenoff and Zimm 1973). In macronuclei of *Tetrahymena* no chromosomes can be detected at any time. Only Seyffert (1979) assumes the existence of chromosome complexes (see discussion in Sect. 4.1.4). Preer and Preer (1979) determined the size of macronuclear DNA molecules by sedimentation velocity analysis in sucrose gradients. They found that the size of the majority of macronuclear DNA molecules was in the range of about 6.5×10^2 kilo base pairs. Merkulova and Borchsenius (1976) reported a similar value. As a consequence of this relatively low size, one has to assume that each micronuclear chromosome is divided into about 70 fragments on the average during macronuclear development.

Williams et al. (1978) found with viscoelastometric measurements a size of 3–4.5×10^4 kilo base pairs for macronuclear DNA of *T. pyriformis* and *T. thermophila*. This value is in the range that can be expected for chromosome-sized molecules. But this value might be skewed toward the largest molecules, since viscoelastometry preferentially measures the largest molecules rather than the average-sized as Preer argues. On the other hand, Preer's measurements might be incorrect due to some breakage of large DNA molecules during the analysis or due to anomalous sedimentation behavior of such large molecules in sucrose gradients. Obviously, the discrepancy between these two measurements cannot be definitely resolved without further experiments. Perhaps another experimental direction might circumvent the technical problems of measuring these large molecules by conventional methods. For instance, titration of free macronuclear DNA ends by means of hybridization with an end-specific labeled hybridization probe which is now available (see Sect. 4.1.5.1) would be an alternative.

However, both measurements clearly show that macronuclear DNA of *T. thermophila* and *T. pyriformis* does not contain large amounts of short gene-sized DNA molecules as they occur in macronuclei of hypotrichous ciliates (Sect. 5.3.1). The size of the DNA molecules is also considerably larger than in the closely related species *Glaucoma chattoni* (Sect. 4.2). Finally, no evidence for covalently linked super-chromosomal aggregates (Sammelchromosomen) could be detected.

4.1.3 Differences in Chromatin Structure and Histone Composition Between Macro- and Micronuclei

Investigations of chromatin structure and histone composition were performed mainly with *T. thermophila*, only very few data have been published for other species (e.g., Hamana and Iwai 1971; Johmann and Gorovsky 1976 b). A review of earlier investigations has been given by Gorovsky et al. (1978).

In electron microscopic spreads of macro- and of micronuclear chromatin of *Tetrahymena*, typical nucleosomes are visible. Brief digestion with staphylococcal nuclease results in a shorter repeat length (175 bp) in micronuclear chromatin compared to macronuclear chromatin (202 bp) (Gorovsky and Keevert 1975; Gorovsky et al. 1978). A similar difference in repeat length was found between macro- and micronuclear chromatin of hypotricous ciliates (see Sect. 5.3.4).

In addition to the small differences in chromatin structure of active and inactive genomes, significant differences have been found in histone composition. Macronuclei contain five major histone fractions (H1, H2A, H2B, H3, and H4). Some of these fractions show a high heterogeneity presumably mainly due to methylation and phosphorylation (Hamana and Iwai 1971; Johmann and Gorovsky 1976a; Gorovsky et al. 1974). Only H4 comigrates with calf thymus H4 histone (Gorovsky 1973). One very heterogeneous fraction, formerly called HX, could be shown to be homologous with H2A only after extensive analytical effort (Johmann and Gorovsky 1976a; Glover and Gorovsky 1978). While macronuclear histone composition is compatible to that found in other eukaryotic organisms, micronuclei contain a considerably deviating set of histones. In earlier publications micronuclei were assumed to lack histones H1 and H3 completely (Gorovsky and Keevert 1975; Johmann and Gorovsky 1976a; Gorovsky et al. 1978). But later studies showed that in the previous investigations micronuclear histone fractions were artifactually lost during the preparation. In fact, micronuclei contain two specific forms of H3, $H3^f$, and $H3^s$. Whereas $H3^s$ is very similar to macronuclear H3, the micronucleus-specific form $H3^f$ is derived from $H3^s$ by a regulated proteolytic processing event (Allis et al. 1979; Allis et al. 1980a). Micronuclei lack histone H1 of the type found in macronuclei, but contain three other micronucleus-specific fractions of histones (α, β, γ) which seem to be homologous to H1 due to their localizations in the linker region of micronuclear chromatin (Allis et al. 1979, 1980b).

An additional difference between macro- and micronuclear histones was detected among the minor histone components. Macronuclei contain two minor primary sequence variants undetectable in micronuclei (Allis et al. 1980b, 1982). The variant hv2 is a sequence variant of H3, hv1 resembles H2A, but seems to be a distinct minor histone variant. Immunofluorescent studies with antibodies directed against hv1 indicate that hv1 is enriched in nucleoli (Allis et al. 1982). This variant has been detected in other *Tetrahymena* species and seems to be highly conserved in evolution, since the antibodies also react with mammalian cells.

Further evidence for histone rearrangements occurring during nuclear differentiation was presented recently (Allis et al. 1984). Histone synthesis and deposition of the different fractions into the nuclei were monitored. Micronucleus-specific histone fractions ($H3^f$, β, γ) seem to disappear from micronuclei as conjugation proceeds. This might be a prerequisite for macronuclear differentiation, since the new macronucleus descends from a nucleus (the macronucleus anlage, cf. Fig. 2) which is morphologically indistinguishable from micronuclei shortly after division of the synkaryon. Macronucleus-specific histones (H1, hv1, hv2) are added to macronuclear chromatin when the developing anlage becomes morphologically and functionally distinct from micronuclei. Later, the new micronuclei regain the micronucleus-specific histones.

It remains to be demonstrated in further experiments whether the rearrangements in histone composition during nuclear differentiation play an essential role in gene activation.

4.1.4 Organization of Macronuclear Genomes

Any discussion of molecular rearrangements during nuclear differentiation must include precise assumptions on the genome organization of both macro- and micronuclear genomes. The micronuclear genome organization seems to be clear: micronuclei of the species of the *Tetrahymena pyriformis* species complex contain ten chromosomes (2N) and genes seem to be arranged accordingly in linkage groups (Bruns 1984). Less well-founded hypotheses concerning the macronuclear genome organization exist. Several hypotheses have been proposed explaining the genomic organization in macronuclei of *Tetrahymena* (reviewed, e.g., by Raikov 1976, 1982; Nanney 1980; Gorovsky 1980). Any hypothesis of macronuclear organization in *Tetrahymena* must explain or be at least compatible with several enigmatic properties of that genome. The most important genetic phenomenon in this respect is the phenotypic assortment. This phenomenon shall be described briefly (for details see Sonneborn 1974b; Nanney and Preparata 1979; Nanney 1980). The assortment of phenotypes can be correlated to morphological characteristics, enzyme mobilities, or serological types. When a cell clone of *Tetrahymena* is cultivated which is heterozygous for a pair of alleles, phenotypic assortment means that the clone produces subclones by vegetative divisions which express only one allele and other subclones which express only the other allele. This assortment yields a stable condition. A phenotype lost during assortment does not return during the vegetative life of that subclone even though the micronucleus remains heterozygous.

Two interpretations of this genetic phenomenon are possible: (1) phenotypic assortment is based on permanent allelic suppression or (2) it is based on a physical loss of alleles, i.e., an allelic assortment. The second interpretation seems to fit the experimental data best according to most research groups.

Several hypotheses attempt to explain macronuclear genome organization as well as genetic phenomena as phenotypic assortment.

The subnuclear hypothesis assumes that the macronucleus of *Tetrahymena* consists of about 45 haploid subnuclei (e.g., Nanney 1964, 1980). This is an agreement with the macronuclear DNA contant, since the DNA content is too small to contain 45 diploid units (see Sect. 4.1.2). It is also in agreement with about 45 assorting units in phenotypic assortment (Orias and Flacks 1975). But if the assorting units were haploid genomes, linked genes should be assorted in linkage groups. Since no clear evidence of somatic linkage in phenotypic assortment has been presented so far, one has to assume macronuclear elements smaller than whole chromosomes. This assumption would agree with the experimental evidence for subchromosomal-sized macronuclear DNA in *T. thermophila* and *T. pyriformis* (see Sect. 4.1.2). But since these results are controversial and the fragments reported, relatively large (about 650 kbp on the average), one should perhaps wait for experimental evidence of somatic linkage of genes which are in close proximity. Alternatively to the assumption of subchromosomal assorting elements, one could supplement the hypothesis with the postulate of a high rate of somatic recombination. This idea is currently under investigation.

Another hypothesis regarding macronuclear structure in *Tetrahymena*, the "nucleosomal" hypothesis, was proposed by Vorob'ev et al. (1975). Unfortunately, the term "nucleosomal" is completely misleading. The term "nucleosome" now generally used for a defined basic structure of the chromatin is used by the authors cited for supposed genome fragments which contain one or several genes. During macronuclear development these presumptive nucleosomes are assumed to be somatically paired. That means that each nucleosome always contains both alleles of the diploid genome. The macronucleus contains after having finished the development 22–23 diploid nucleosomes. Afterwards the diploid nucleosomes split into two haploid ones (allelic splitting). Thus, approximately 45 assorting units would result. After replication haploid nucleosomes may segregate independently and at random. When by random segregation all nucleosomes with one of two alleles are lost by chance the phenomenon of phenotypic assortment which is based on an allelic assortment would result. The main problem connected with this hypothesis is unresolved. How is it possible that clones of *Tetrahymena* live for many years vegetatively without renewal of their macronuclei and obviously without losing both alleles? Such a loss should produce a high percentage of defective cells with in fact are not observed. Additionally, there is no cytological evidence for 22–23 diploid nucleosomes or for the postulated "allelic splitting". Also this hypothesis demands a relatively large amount of fragmentation of the anlagen genome during development.

An interesting alternative assumption on macronuclear organization of *Tetrahymena* has been proposed by Seyffert (1979). It is based on cytophotometric measurements of chromatin structures in spreads of *Tetrahymena* macronuclei. Seyffert detected five groups of chromatin substructures in spread macronuclei. The DNA contents of these substructures correspond to multiples of the basic values of micronuclear chromosomes. The structures were interpreted as to consist of bundles of chromosomes. Following this interpretation the distributional units in dividing macronuclei were multiples of individual chromosomes. Also this hypothesis poses some severe problems as Seyffert admits. The "amitotic" division of macronuclei demands regulatory mechanism which must prevent a complete loss of one type of chromosome which otherwise should occur rapidly.

Perhaps an indication for the existence of such a postulated specific regulatory mechanism might be seen in the following observations. The macronuclear DNA content can be regulated in vegetative cells of *Tetrahymena* (for review see Doerder 1979). Extra S-phases and the generation of chromatin extrusion bodies seem to fulfill that function (e.g., Cleffmann 1968). But so far it has not been tested whether these mechanisms provide only a coarse control of the whole macronuclear DNA content or whether they are able to compensate for gene imbalances. An additional problem arises. DNA elimination, chromosome breakage, rearrangements (see Sect. 4.1.5), as well as phenotypic assortment, are difficult to explain in the light of the last hypothesis.

The conclusion must be drawn that so far no hypothesis regarding macronuclear genome organization in *Tetrahymena* is available which allows a conclusive explanation for all experimental data.

4.1.5 Amplification, Elimination, and Rearrangement of Individual Sequences

In the previous section the bulk differences between macro- and micronuclear genomes of *Tetrahymena* have been considered. In this section alterations at the level of individual sequences shall be discussed. These alterations shall give a deeper insight into the molecular mechanisms which may finally result in the observed bulk differences with regard to DNA related characteristics. Most of these more detailed comparisons were only possible after the development of gene cloning methods.

4.1.5.1 Ribosomal RNA Genes

The most available information is on developmental changes of the genes coding for ribosomal RNA (rDNA). Initially, the basic studies could be done without using cloned DA fragments due to the particular structure and multiplicity of these genes in *Tetrahymena* macronuclei. Since reviews dealing with rDNA structure and amplifications have been published by several authors (Gall et al. 1979; Yao et al. 1979; Yao 1982; Blackburn 1982), only the most important data, which are essential for further discussion shall be described here.

In micronuclei of *T. thermophila* only a single rDNA copy per haploid genome is present (Yao et al. 1974; Yao and Gall 1977). It has been localized in chromosome 2 of *T. thermophila* as indicated by Southern blot hybridizations to DNA isolated from different nullisomic strains (Yao 1982 b). Nullisomic cells are lacking both copies of one or more of the five chromosomes in the micronucleus (Bruns and Brussard 1981). The micronuclear rDNA copy is integrated into high molecular weight chromosomal DNA. This was shown by hybridization of labeled cRNA transcribed from purified macronuclear rDNA to Southern blots of restriction enzyme-digested micronuclear DNA (Yao and Gall 1977).

In contrast to the situation found in *Tetrahymena* micronuclei, the macronuclei obviously do not contain integrated rDNA copies. About 10,000 extrachromosomal rDNA copies per macronucleus (in G1) are found (Engberg et al. 1974; Yao and Gall 1977). Each macronuclear rDNA molecule is a large palindrome about 21 kb in length consisting of two rRNA transcription units which are connected in a inverse repeat orientation (see Fig. 4) (Gall 1974; Karrer and Gall 1975, 1976; Engberg et al. 1976). The generation of extrachromosomal rDNA during macronuclear development involves rDNA excision from the micronuclear chromosome, reorganization of the excised copies, and amplification (reviewed by Yao 1982 b). The molecular mechanisms of the generation of macronuclear rDNA are not fully understood, but a considerable number of important details are known:

1) The micronuclear rDNA copy is cut out of the chromosomal DNA by chromosome breakage (Yao 1981; Pan and Blackburn 1981) shortly after the beginning of the anlagen development.

2) A DNA fragment of about 2.8 kb adjacent to the integrated micronuclear rDNA is eliminated from the developing anlage (Yao 1981).

3) The excised rDNA must be multiplied by replication and the resulting copies must be connected in a head-to-head orientation by a hitherto unknown mechanism. Intermediates of the amplification process have been found (Pan and Blackburn 1981; Pan et al. 1982) in developing anlagen. They consist of nonpalindromic molecules 11 kb in length which represent one-half of the final palindromes as proven by restriction enzyme mapping and hybridization experiments.

4) The dimers are amplified by independent replication rounds to give the final rDNA copy number. The true mechanism is unclear. It might be that not the dimers, but the monomers are amplified and connected afterward.

5) Concomitant with excision and amplification, specific telomere structures are formed at the ends of the dimers.

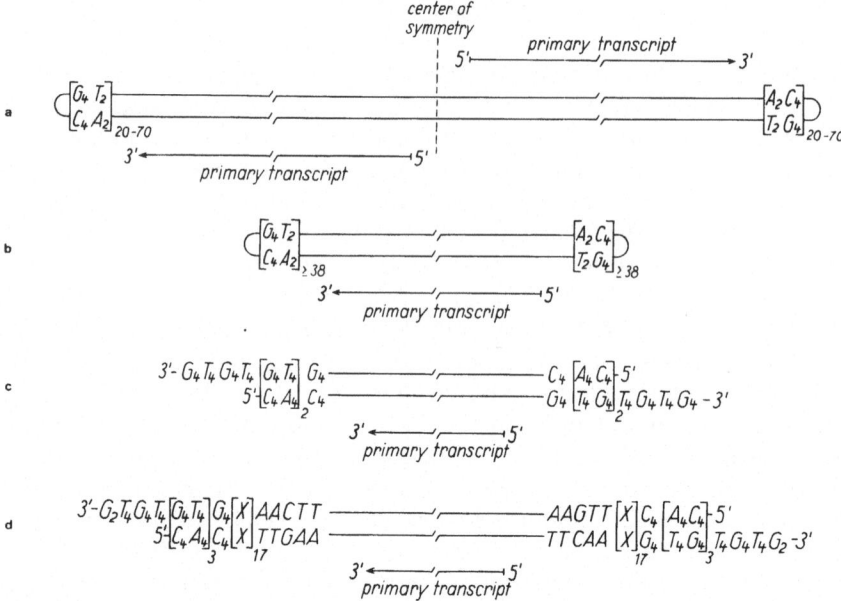

Fig. 4 a–d. Free, linear macronuclear genes of ciliates (for references see text). **a** rDNA of *Tetrahymena*. The palindromic dimers of about 20 kb contain two transcription units. The ends of the molecules consisting of C_4A_2 repeats have a covalently closed hairpin structure with several single-nucleotide gaps (not shown in the Fig.) near the ends. **b** rDNA of *Glaucoma*. Monomers of 9.3 kb containing one transcription unit. Ends of the molecules as in *Tetrahymena*. The protein-coding genes of *Glaucoma* have presumably the same structure, but individual genes except the ribosomal genes have not been investigated so far. **c** General structure of all macronuclear genes of hypotrichous species of the genera *Stylonychia*, *Oxytricha*, *Urostyla*, and *Paraurostyla*. Macronuclear DNA molecules range in size from 0.3 kb to about 20 kb. Every DNA molecule contains only one transcription unit according to present knowledge. **d** General structure of macronuclear genes of the hypotrichous species *Euplotes aediculatus*. A slight difference in the terminal repeat sequence can be seen if compared with macronuclear genes of other hypotrichous species. It seems doubtful that the termini of the macronuclear DNA molecules of the hypotrichous species have the structure as shown here also in vivo. It is possible that terminal hairpins or single-stranded complementary RNA or DNA fragments, attached to the single-stranded extension by hydrogen bonds, have been lost during isolation

Telomeres are specific structures found at the ends of linear DNA molecules or at chromosome ends (for reviews see Blackburn et al. 1983; Blackburn and Szostak 1984; Blackburn 1984). Several functions of the telomere structures are discussed. Among others they may allow the replication of DNA ends and/or mediate the connection between DNA molecules and proteins of the nuclear matrix or nuclear envelope.

The ends of the macronuclear rDNA molecules carry 20–70 tandem repeats of the hexanucleotide CCCCAA (Blackburn and Gall 1978; Yao and Yao 1981). The very last end is formed by a terminal, covalently closed hairpin loop (see Fig. 4). Within the terminal repeats sequence-specific single-nucleotide gaps exist. Instead of tandem repeats, only a single CCCCAA block is found adjacent to both sides of the integrated rDNA copy of the micronucleus (King and Yao

1982). The interesting question whether the terminal repeats are added step-by-step to the ends of the extrachromosomal rDNA molecules or by a recombination process cannot as yet be answered (Blackburn et al. 1983). The function of the rDNA telomeres could be demonstrated in a transformation experiment (Szostak and Blackburn 1982). Terminal restriction fragments of *Tetrahymena* rDNA were isolated and ligated to both ends of a linear plasmid of yeast. The recombinant linear plasmid containing the termini of *Tetrahymena* rDNA was shown to be replicated and maintained in yeast cells after transformation. A hairpin structure alone was insufficient in fulfilling that function.

Also the organization of 5S-rRNA genes in macro- and micronuclei of *Tetrahymena* has been studied (Toennesen et al. 1976; Kimmel and Gorovsky 1976). No differences could be detected in hybridization experiments. But more recent investigations reveal that some 5S-rRNA clusters are micronucleus-specific and others macronucleus-specific. The total number of 5S-rRNA gene copies per haploid genome is about 33% greater in macro- than in micronuclei.

4.1.5.2 Protein Coding Genes

So far only two protein coding genes of known function have been studied under the aspect of putative rearrangements during nuclear differentiation. Callahan et al. (1984) reported that macro- and micronuclei of *T. thermophila* contain a single type of expressed α-tubulin gene. Whereas the transcribed sequence seems to be identical in macro- and micronuclei, indications of rearrangements of the flanking sequences were found. Hybridizations to Southern blots of digested macro- and micronuclear DNA suggested that breakage and rejoining in the flanking regions must occur during nuclear development. Sequence analysis must be awaited for a deeper understanding.

A similar study was done with histone H4 genes in *T. thermophila* (Bannon et al. 1984). Two types of H4 histone genes were found in Southern blot analysis of macro- and micronuclear genomes. But no indication of rearrangements either of the transcribed or of the adjacent flanking sequences could be detected. But the question whether very small deletions or rearrangements comparable to those found by Klobutcher et al. (1984) in *Oxytricha* would have been detected in these experiments remains open.

4.1.5.3 Other Sequences

Several less well-defined sequences were studied with respect to their micro- and macronuclear organization. Randomly chosen sequences belonging to the repetitive as well as to the unique fraction of the micronuclear genome were cloned in plasmid or phage vectors. Their arrangement and distribution was probed in hybridization reactions with micro- and macronuclear DNA (Iwamura et al. 1982; Brunk et al. 1982; Yao 1982a; Yokoyama and Yao 1982; Karrer 1983; Yao et al. 1984; Diamond et al. 1984).

In situ hybridizations and hybridizations with micronuclear DNA of nullisomic strains (see Sect. 4.1.5.1) suggested that germ line specific repetitive sequences are present on all five micronuclear chromosomes (Yao 1982a;

Yokoyama and Yao 1982; Karrer 1983). Extensive rearrangements could be observed with members of families of repetitive sequences, whereas only a few rearrangements were detected with nonrepetitive sequences (l.c.).

With regards to mechanisms of elimination, the interesting result was obtained by quantitative in situ hybridizations that micronucleus-specific sequences seem first to be replicated a few times before they were eliminated from the anlage (Yokoyama and Yao 1982). This seems especially interesting in comparison to replication and elimination in macronuclear development of hypotrichous ciliates (Sect. 5.2). An indication as to the mechanism of discrimination between sequences which are eliminated and those which are retained was obtained by White and Allen (1984). A member of a family of repetitive sequences which could be shown to contain a site of specific methylation was obviously not eliminated in contrast to all other nonmethylated members of that family.

Recently recombinant clones could be obtained from micronuclear DNA which contained the junctions between eliminated and maintained sequences. The surprising result was obtained that the elimination of most sequences is coupled with a rejoining of the remaining sequences (Yao et al. 1984). It was calculated from the data that more than 5,000 such rearrangements may occur in the *Tetrahymena* genome. This mechanism of breakage and rejoining, a true arrangement which resembles the rearrangement of immunoglobulin genes in mammals (reviewed by Tonegawa 1983) seems to differ from the simpler fragmentation process which leads to the generation of rDNA (see Sect. 4.1.5.1).

Furthermore, the organization of the C_4A_2 repeats is different between micro- and macronuclei. In macronuclear DNA, clusters of C_4A_2 repeats are found not only at the ends of the rDNA, but in many other places (Yao et al. 1981). They seem to be always located near free ends as indicated by Bal 31 sensitivity. In micronuclear DNA about 300 C_4A_2 repeats were found integrated into longer DNA molecules. They were not digestable with Bal 31 (Yao et al. 1981; Yao and Yao 1981).

4.1.6 Putative Somatic Functions of Micronucleus-Specific Sequences

Micronuclei of *Tetrahymena* are generally assumed to be transcriptionally inactive nuclei as in other ciliates, too (see Chap. 2). But there is some contradictory evidence regarding this important aspect. A large percentage of *Tetrahymena* strains isolated from the wild is amicronucleate, but viable for many generations, e.g., *T. pyriformis* strain GL (see Sect. 4.1.1). On the other hand, cells from which micronuclei have been removed, for instance, by irradiation (e.g., Wells 1960) or by other means, are generally not viable. These experiments lead to the conclusion that micronuclei of *Tetrahymena* contribute essential information to vegetatively growing cells. But the existence of viable amicronucleate strains provide evidence to the contrary.

This problem raised the question as to the occurrence and synthesis of RNA in *Tetrahymena* micronuclei. Gorovsky and Woodard (1969) did not detect RNA or RNA synthesis in micronuclei of *Tetrahymena* after analyzing ^{3}H-uridine incorporation by means of autoradiography. But Murti and Prescott (1970) and Sugai and Hiwatashi (1974) presented evidence for micronuclear RNA synthesis.

Sugai and Hiwatashi showed clearly by using pulse labeling and chase experiments that during the crescent stage of the meiotic prophase (leptotene and zygotene) micronuclei are actively synthesizing RNA. The RNA synthesis found by these authors resembles the RNA synthesis on lampbrush chromosomes during meiotic prophase of amphibian oocytes. Transcription of micronuclear DNA during meiosis might pertain to conjugation events, but it is rather unlikely that it influences vegetative functions of the cells significantly.

Murti and Prescott (1970) detected RNA synthesis during micronuclear S-phase in vegetatively growing cells. They used somewhat more sophisticated techniques compared to Gorovsky and Woodard (1969), since they labeled well synchronized cells and evaluated the uridine incorporation by electron microscopic autoradiography.

In both cases of putative micronuclear RNA synthesis, it remained uncertain as to whether the RNA is degraded within the micronuclei or transferred into the cytoplasm. In addition, the possibility exists that newly synthesized RNA of macronuclear origin was transferred into micronuclei and, therefore, not synthesized in situ.

Recently exciting progress was made by Karrer et al. (1984) regarding the problem of putative somatic functions of the micronuclear genome. The authors used cloned sequences which had been shown to be micronucleus-specific (see Sect. 4.1.5.3) for in situ hybridizations and hybridizations to Southern blots. Macronuclei and macronuclear DNA were isolated from amicronucleate and from wild-type micronuclei-containing strains of *T. thermophila*. The amicronucleate strain was isolated from cells which had been mutagenized with nitrosoguanidin by Kaney and Speare (1983). As an exception to the rule (see above), this artifactually amicronucleated strain is viable. The hybridization experiments gave the surprising result that the micronucleus-specific sequences hybridized strongly with the macronuclear DNA of the amicronucleate strain, but not with that of normal strains. It was concluded that incomplete elimination of micronucleus-specific sequences during macronuclear development might be correlated with the viability of amicronucleate strains of *T. thermophila*. The sequence did not react with macronuclear DNA of an amicronucleate strain of the species *T. pyriformis*. Thus, there are perhaps differences between species with respect to the eliminated sequences.

These experiments provide an interesting approach to answering the question whether micronuclei have in addition to their germ line functions other essential functions in the vegetative life of the cells. Whether such essential micronuclear sequences may act by means of transcription or by other unknown mechanisms remains open to speculation.

4.1.7 Concluding Remarks

Nuclear differentiation in *Tetrahymena* species exhibits considerable molecular reorganization both at the level of nuclear proteins as well as in DNA sequence structure. Differences of about 10–20% between macro- and micronuclear genomes are caused by several distinct molecular mechanisms. Amplification concerns mainly 17S and 25S ribosomal RNA coding genes, perhaps to a lesser extent also 5S RNA genes. Mainly members of repeat families are eliminated from developing macronucleus anlagen. True rearrangements in form of breakage and rejoining seem to be a common phenomenon in nuclear development. Whereas transcribed sequence regions seem to be maintained unchanged during development, the flanking regions are in some, but perhaps not in all, cases involved in rearrangements.

4.2 *Glaucoma*

The ciliate species *Glaucoma chattoni* is closely related to *Tetrahymena* species. It belongs to the same suborder Tetrahymenina of the Hymenostomatida (Corliss 1979). But the organization of the macronuclear genome of *Glaucoma* exhibits very surprising differences compared to that of the *Tetrahymena* species. Micronuclear DNA of *G. chattoni* has not been investigated so far, but it is organized in chromosomes as in *Tetrahymena* micronuclei and may also be assumed to be similar in other features. But macronuclear DNA has a considerably lower molecular weight than that of *Tetrahymena* macronuclei (Sect. 4.1.4) ranging in size from above 100 kb down to 2.1 kb as measured by sedimentation centrifugation, gel electrophoresis, and electron microscopy (Katzen et al. 1981). A significant fraction of the macronuclear genome exists as discrete-sized classes. Among them a class of 9.3 kb molecules contains the ribosomal RNA genes. No rDNA dimers are found as in *Tetrahymena*, each linear rDNA molecule of *Glaucoma* carries a single coding region (cf. Sect. 4.1.5.1). Presumably all subchromosomal macronuclear DNA molecules have tandem C_4A_2 repeats in inverted orientation at their ends (see Fig. 4). The structure of the DNA termini was analyzed by sequencing and by electron microscopic analysis after denaturation and renaturation. Upon denaturation and quick renaturation single-stranded circles with short, double-stranded "panhandles" are formed due to the inverted repeats at the termini. This was shown earlier for macronuclear DNA of hypotrichous ciliates (Wesley 1975, cf. Sect. 5.3.2).

Thus, on the whole, the organization of the macronuclear genome of *G. chattoni* resembles that found in hypotrichous ciliates more than that of the closely related *Tetrahymena* species. Macronuclear development in *Glaucoma* exhibits a considerably higher degree of genom fragmentation than that of *Tetrahymena*. Whether the extent of elimination also reaches the values found in the hypotrichous species can only be detected in comparing macro- with micronuclear DNA of *Glaucoma*. The small size and low DNA content of *Glaucoma* micronuclei makes this difficult.

Undoubtedly these experimental results show that nuclear differentiation can vary considerably even among closely related ciliate species.

4.3 *Paramecium*

Paramecium species belong to the same order, Hymenostomatida, as *Tetrahymena* and *Glaucoma*. Some of them are also well-known unicellular laboratory animals and useful subjects in many areas of biological research (Sonneborn

1974a; Nanney 1964, 1980). But whereas nuclear differentiation has been investigated at the molecular level quite well in *Tetrahymena*, similar studies are almost completely lacking for the *Paramecium* species. Perhaps the general assumption that the macronuclei of *Paramecium* are exact polyploid editions of the micronuclear genomes (Sonneborn 1974a; Raikov 1969) has prevented more extensive research activities in this field. Technical problems have provided additional obstacles. Micronuclei of the most intensively studied species are very small in size and it is difficult to obtain larger quantities of pure, undegraded micronuclear DNA (Cummings and Tait 1975). But it should be mentioned that similar problems have not prevented the successful investigation of micronuclear DNA of *Tetrahymena* (see Sect. 4.1.2).

A few facts regarding macronuclear development, genome organization, and function in *Paramecium* species shall be described here. Some of these data have raised some doubt as to the general validity of the assumption that the macronuclei are exact polyploid editions of the micronuclei. These facts must be considered in any generalization on nuclear differentiation in ciliates as a whole.

4.3.1 The Species Problem

A similar problem exists with species designations in *Paramecium* as in *Tetrahymena* (cf. Sect. 4.1.1) when comparing older literature with the more recent. Most of the earlier work dealt with the "species" *Paramecium aurelia*. It was known that this species consisted of several morphologically indistinguishable, but genetically isolated syngens (for discussion see Sonneborn 1957). When biochemical methods were available for the identification of the syngens without using living reference strains (Tait 1970), the syngens were given true Latin species names (Sonneborn 1975). From this group of closely related species mainly *P. primaurelia* and *P. tetraurelia* are now used for molecular investigations.

4.3.2 Indications of Molecular Reorganizations
During Macronuclear Development

Data concerning the macronuclear genome organization of *Paramecium primaurelia* have been published which might indicate the existence of substantial differences between macro- and micronuclear genomes. McTavish and Sommerville (1980) found that the size of the majority of macronuclear DNA molecules ranges from 0.6–15 kb. This would be a very similar size range as that found for the gene-sized macronuclear DNA molecules of hypotrichous ciliates (Sect. 5.3.1). Unfortunately, the low molecular weight macronuclear DNA of *P. primaurelia* does not show any specific banding pattern when separated in agarose gels. No hybridization experiments were done with single gene probes. Therefore, the possibility of some degradation cannot be excluded with certainty. Careful isolation of macronuclear DNA of the related species *P. tetraurelia* and *P. bursaria* showed that the majority of the DNA molecules were larger than 20 kb (Steinbrück et al. 1981). But it might be that even among closely related species of the *P. aurelia* species complex large differences exist in the process of nuclear differentiation. The differences found between *Tetrahymena* and *Glaucoma* genome organization (cf. Sects. 4.1 and 4.2) would favor this view. In any case, a reinvestigation of the macronuclear genome organization of *P. primaurelia* is highly desirable.

Clear evidence for molecular reorganizations during macronuclear development was presented for *P. bursaria*. Schwartz and Meister (1975) showed with microspectrophotometric measurements that during the anlagen development of *P. bursaria* DNA elimination occurs. After a short increase in DNA content at the beginning of the anlagen development a significant decrease occurs during the "Feulgen-negative" phase. Later, the remaining DNA is replicated until the DNA content of the mature macronucleus is reached. This observation has been supported by cytological observations (Schwartz 1958). During the anlagen development only short parts of the chromosomes are amplified and later incorporated into the new macronucleus. The rest of the chromosomes seems to be eliminated. Obviously, this elimination and amplification process varies between different strains of *P. bursaria*.

An investigation of the corresponding phases in macronuclear development of a *P. aurelia* species (Berger 1973) did not provide any evidence for DNA elimination in those species. But perhaps the measurements should be repeated and done in shorter time intervals during development and with a more sensitive technique, e.g., cytofluorometry.

There are further observations of existing and putative differences arising during macronuclear development in *Paramecium*. Ribosomal RNA genes occur as extrachromosomal copies in macronuclei of *P. tetraurelia* (Findly and Gall 1978, 1980). But in contrast to *Tetrahymena* (Sect. 4.1.5.1), no palindromes are found. The rDNA consists of linear and circular molecules with a varying number of tandem repeats (from 2 to more than 13) as indicated by electron microscopic analysis. The organization of micronuclear rDNA has not been analyzed. But since the extrachromosomal macronuclear rDNA exhibits intramolecular heterogeneity in the tandem repeats, the authors assume that more than one repeat must exist in micronuclei. Thus, despite the difference in rDNA organization of *P. tetraurelia* to that of *Tetrahymena*, one has to assume that also in this case, a gene amplification and perhaps excision mechanism is acting during nuclear differentiation.

The detection of C_4A_2 repeats located near free DNA ends in *P. tetraurelia* macronuclear DNA (Yao and Yao 1981) is another rather indirect evidence that DNA fragmentation and/or elimination might be associated with nuclear differentiation in a similar way as in *Tetrahymena* (see Sect. 4.1.5).

Recently, an exciting study was presented by Epstein and Forney (1984) on expression of genes coding for surface proteins. This study adds a new aspect to the problem of nuclear differentiation in *Paramecium*. The authors present evidence for the existence of modifier genes in *P. tetraurelia* which affect the expression of immobilization antigen genes. The immobilization antigens (i-antigens) of *Paramecium* are glycoproteins designated A, B, C, etc. which cover the outer surface, including cilia of *Paramecium* cells (for reviews see Preer 1969; Sommerville 1970; Sonneborn 1974a). Usually each cell carries only one type of i-antigen at a time on its surface. These surface proteins are called immobilization antigens since antiserum directed against one i-antigen immobilizes those cells which carry that i-antigen. Thus, cells carrying, for instance, i-antigen A can be characterized by their specific agglutination reaction with anti-A antiserum and are, therefore, called serotype A. A change of environmental conditions can lead to serotype transformation. Transformation of serotypes is reversible and seems to be due to changes in gene expression. Structure and expression of genes coding for i-antigens have been studied by Forney et al. (1983). In the study of Epstein and Forney (1984) a mutant was detected which results in the inability of the mutant cells to transmit the micronuclear copy of the gene coding for i-antigen A to the macronucleus during development. The cells of that mutant do not express serotype A since an intact copy of that gene is present only in the apparently transcriptionally inactive micronucleus and is absent from the transcriptionally active macronucleus. The authors conclude from their results that the *Paramecium* cells possess the ability to control the expression of a gene by regulating its incorporation into the developing macronucleus. At the moment this conclusion remains speculation. It has also been speculated that a similar mechanism might be used for mating type determination in *Paramecium*. On the basis of theoretical considerations, Orias (1981) proposed a similar idea for mating type determination in *Tetrahymena*.

4.3.3 Putative Somatic Functions of Micronuclei

As described for *Tetrahymena* also in *Paramecium* species only very few indications have been published that micronuclei might play a role during the vegetative life of the cells. Pasternak (1967) found transcriptional activity in micronuclei of species belonging to the *P. aurelia* complex. Pulse labeling experiments followed by autoradiography gave clear evidence that in micronuclei of normal cells as well as of cells of an amacronucleate mutant RNA synthesis existed in situ. Rao and Prescott (1967) detected only during micronuclear S-phase some transcriptional activity in micronuclei of *P. caudatum*.

In a more recent investigation Fujishima and Watanabe (1981) could show that abnormalities of amicronucleate cells of *P. caudatum* could be cured by reimplantation of micronuclei. The removal of micronuclei led to prolongation of the cell cycle, decrease in food vacuole formation, and shortening of the buccal cavity in the amicronucleate cells. After reimplantation of micronuclei all these characteristics returned to normal. The authors concluded from their micronuclei transplantation experiments that micronuclei play an important role not only during the sexual cycle, but also in the vegetative life of *P. caudatum*.

4.3.4 Conclusions

Molecular reorganizations occur during macronuclear development at least in some *Paramecium* species. Wether these are exceptions to the rule, should be examined in more species. It seems especially necessary to compare macro- and micronuclear genes of *Paramecium* species by detailed sequence analysis since the differences might be small. Also in *Paramecium* species micronuclei seem to have in addition to their well-known germ line function at least a few somatic functions.

The general assumption that the macronucleus of *Paramecium* species is a polyploid version of the micronucleus does no longer hold true.

5 Nuclear Differentiation in Hypotrichous Ciliates

The most dramatic molecular reorganizations were detected in *Stylonychia* and *Oxytricha* species (see Fig. 1) which belong to the order Hypotrichida. Some molecular data are available on other hypotrichous species of the genera *Euplotes*, *Paraurostyla*, and *Urostyla*.

5.1 The Species Problem

The identification of some hypotrichous ciliate species is extremely difficult. Therefore, considerable confusion exists in species designations even of the better known species. Renaming of species as it was described for *Tetrahymena* and *Paramecium* (Sects. 4.1.1 and 4.3.1) enhances the confusion. Therefore some clarifying remarks are inevitable.

The great interest in molecular events occurring during nuclear differentiation in hypotrichous species began with Ammermann's observation (Ammermann 1965) that a dramatic DNA elimination and the unexpected formation of giant chromosomes characterize macronuclear development in the "species" *Stylonychia mytilus*. Two varieties of *S. mytilus* were known, which differed in size and macronuclear DNA content among other characteristics. They seem to be genetically isolated. Recently, these two varieties were described as separate species since they can be identified on the basis of morphological as well as distinct molecular characteristics (Ammermann and Schlegel 1983; Steinbrück and Schlegel 1983). Unfortunately, that variety which was used in the vast majority of earlier investigations was given the new name *S. lemnae* due to the rules of zoological nomenclature. Therefore, earlier experimental results (until 1983) obtained with *S. mytilus* variety 1 and more recent data gained from *S. lemnae* relate to the same species.

But there are more complications. The former *S. mytilus* variety 1 (now *S. lemnae*) was brought by Ammermann to the laboratory of Prescott in Boulder, Colorado, U.S.A. Many important contributions regarding nuclear differentiation were published by Prescott's research group (see below) until the original species died out in the laboratory. New animals were collected from the wild. Initially, these new clones were believed to belong to the same formerly used species. But a reexamination revealed that a different species has been collected from the wild which belonged to the closely related genus, *Oxytricha* (see Lauth et al. 1976). Therefore, some data published before 1976 may in fact relate to an *Oxytricha* species and not to *Stylonychia*. Macronuclear development does not differ significantly between *Oxytricha* and *Stylonychia* species, but differences are found in details.

Several *Oxytricha* species have been used for molecular investigations, *O. fallax*, *O. similis*, and *O. "nova"*. *O. "nova"* is a laboratory name only. According to Schlegel (1985), *O. "nova"* is identical with *O. bifaria* as indicated by isoenzyme analysis. In addition, careful systematic analyses have raised severe doubts whether the two genera *Oxytricha* and *Stylonychia* can be separated from each other reproducibly. Presumably only one genus exists (Schlegel 1985). Accordingly, further extensive renaming must be awaited. Without exaggeration one may conclude that systematic and species delimitation of the Oxytrichidae is in a predicament.

5.2 Peculiarities of the Macronuclear Development in Hypotrichs

Macronuclear development is a somewhat more complex process in hypotrichs than in other ciliates (see Chap. 2 for basic events and cf. Fig. 2). It has been investigated in *Stylonychia*, *Oxytricha*, and *Euplotes* species in detail (Ammermann 1965, 1971; Ammermann et al. 1974; Spear and Lauth 1976). The whole process will be described for *S. lemnae* (formerly *S. mytilus*, variety 1, see Sect. 5.1) as an example. Differences are mentioned briefly. The thereby arising important molecular differences are discussed later (Sect. 5.3).

S. lemnae performs conjugation in a very similar way as described above for ciliates in general (Chap. 2). Figure 3a shows a pair of conjugating cells of *S. lemnae*. But after separation of the exconjugant cells, a characteristic series of cytological events starts (see Fig. 5). After mitotic division of the synkaryon the prospective macronucleus anlage enlarges and in diploid exconjugants about 300 chromosomes become visible. Two-thirds of these chromosomes become pycnotic and are expelled into the cytoplasm. The DNA content of the anlage is at this early stage of development already larger than that of a diploid micronucleus. Furthermore, the eliminated chromosomes are presumably already oligotenic as based on microspectrophotometric measurements (Ammermann, pers. commun.). This step of chromosome elimination is not visible in Fig. 5 as a drop in the DNA content of the early anlage, since the polytenization of the maintained chromosomes compensates the DNA loss caused by the elimination of two-thirds of the chromosomes.

In *Oxytricha* obviously all chromosomes become polytenic (Spear and Lauth 1976). Also for *Euplotes* species no stage of chromosome elimination has been described so far (Rao and Ammermann 1970). Perhaps weak evidence for a chromosome elimination step occurring during macronuclear development of *E. aediculatus* might be found in an observation of Zier (1972). She detected a lower number of free chromosome ends in *Euplotes* anlagen in the giant chromosome stage as it should be supposed from the chromosome number of the diploid chromosome set of the micronuclei of this species. The genetic problems and putative functions of this step of chromosome elimination will be discussed later (see Chap. 8).

After presumably five rounds of replication the anlage of *S. lemnae* has reached its maximal size (Ammermann et al. 1974). An exconjugant of *S. lemnae* containing a large anlage in the giant chromosome stage is shown in Fig. 3b. In *S. lemnae* the giant chromosomes differ characteristically in their appearance from giant chromosomes as they are found, for instance, in *Drosophila* or *Chironomus* species. They contain large, so-called heterochromatic blocks (see Fig. 6) which represent regions which are heavily overreplicated (l.c.). Similar heterochromatic blocks are found in a considerably smaller form in giant chromosomes of *Euplotes* species and are not detectable in giant chromosomes of *Oxy-*

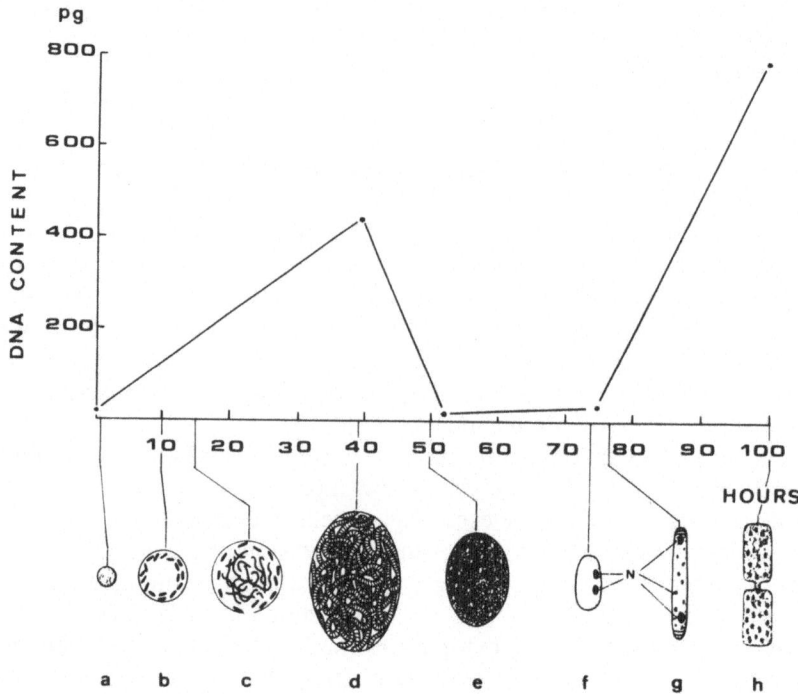

Fig. 5 a–h. Changes of the DNA content during macronuclear development of *Stylonychia lemnae* (modified after Ammermann et al. 1974). **a** After separation of the exconjugant cells (0 h), the young macronucleus anlage has nearly the DNA content and the size of a diploid micronucleus. The increase of the DNA content starts. **b** Later, the anlage swells considerably and small condensed chromosomes become visible situated at the periphery of the anlage. **c** After 15 h about one-third of the chromosomes despiralize and begin to develop into giant chromosomes. The remaining condensed chromosomes disappear and are eliminated into the cytoplasm (chromosome elimination step). **d** The DNA content reaches a first peak after 35–40 h, the giant chromosomes show their maximum size at this time. **e** Giant chromosomes are fragmented, their bands are enclosed into independent vesicles. The DNA content drops drastically, more than 90% of the anlagen DNA is eliminated into the cytoplasm (DNA elimination step). A DNA-poor stage of about 36 h–72 h follows. **f** The end of the DNA-poor stage is marked by the appearance of two nucleoli (*N*) in the anlage. **g** Replication bands become visible and new DNA replication rounds start. The anlage elongates. **h** After five rounds of replication the anlage has reached the DNA content of the final macronucleus

tricha (Ammermann 1971; Spear and Lauth 1976). In contrast to the well-known giant chromosomes of dipteran larvae, the giant chromosomes of hypotrichous ciliates show no transcriptional activity. Contradictory evidence was assumed by some authors to have been presented. Gaude (1981) found RNA polymerase II in giant chromosomes of *Stylonychia* and other authors (Pérez-Silva and Alonso 1966; Jareno et al. 1972) assumed to have detected puffs in giant chromosomes of *Stylonychia*. But neither author could present convincing evidence for newly synthesized RNA which can be removed by RNase digestion. The presence of RNA polymerase might be correlated with the synthesis of RNA primers which are required for DNA replication. The diffuse structures resembling puffs might

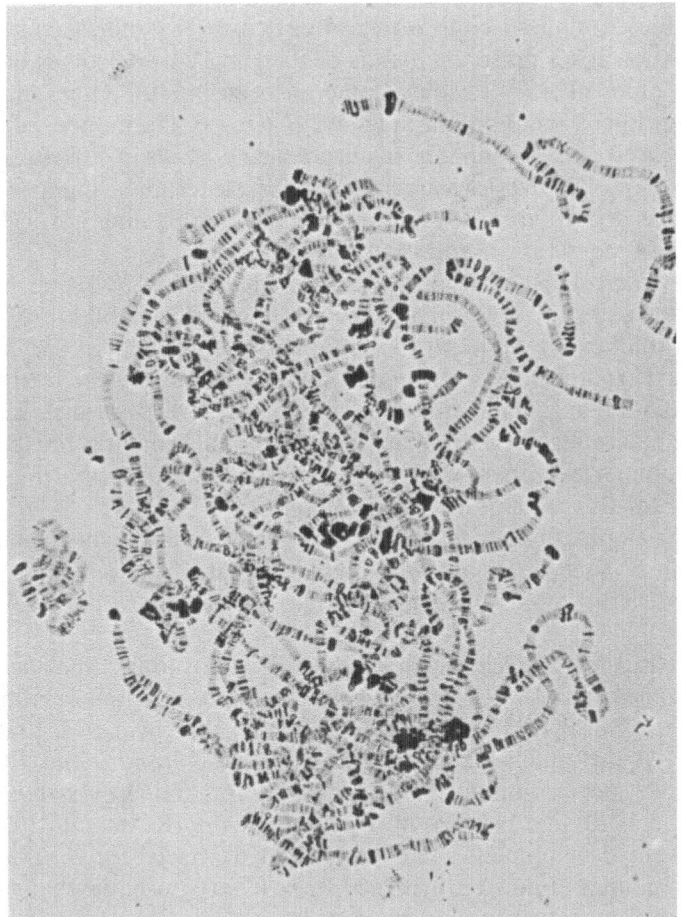

Fig. 6. Giant chromosomes of the macronucleus anlage of *Stylonychia lemnae*. Squash preparation, stained with orcein-lactic acid. Bands and interbands are clearly visible. The dark staining bands – called heterochromatic blocks – are a peculiarity of this species. Courtesy of D. Ammermann

be assumed to be artifacts due to squashing the chromosomes in slightly alkaline medium (personal observation).

After polytenization a second elimination step follows in the anlagen development of *S. lemnae* (see Fig. 5). A drastic decrease of the DNA content of the developing anlage can be seen in Fig. 5. A similar step of the extensive chromatin elimination has been found in all hypotrichs investigated so far. The extent of elimination differs from species to species. About 93% of the anlagen DNA is eliminated in *S. lemnae*, about 80% in *Euplotes aediculatus*, and about 96% in *Oxytricha* sp. (Ammermann 1971 a; Ammermann et al. 1974; Prescott et al. 1979). Thus, taking both elimination steps of *S. lemnae* anlagen development together more than 98% of the micronuclear DNA is eliminated.

The molecular mechanisms of the chromatin elimination process are hitherto unknown. Electron microscopic observations give a few indications. Shortly before elimination takes place, the bands of the giant chromosomes become separated from each other by fibrous septae (Kloetzel 1970). This results in about 10,000 independent vesicles in the anlage of *S. lemnae*. The number of vesicles correlates quite well with the number of chromosome bands in *Stylonychia*. In *Oxytricha* only about 2700 vesicles are counted despite a similar number of chromosome bands as in *Stylonychia*. It is assumed that several bands are enclosed in one vesicle in this species (Spear and Lauth 1976).

Electron microscopy of surface spread anlagen of *Stylonychia* revealed a characteristic loop structure of the giant chromosomes (Meyer and Lipps 1980, 1981). Large lateral loops are attached to 10 nm chromatin fibers in the interband regions of the chromosomes. During chromatin elimination, the lateral loops are released from the axis. The authors assume that the DNA which is maintained in the final macronucleus is localized in the 10 nm chromatin fibers of the axis, whereas the material of the lateral loops is eliminated. But supporting experimental evidence for this assumption is still lacking.

After the elimination stage a DNA-poor stage follows which lasts for about 36 h–72 h in *S. lemnae*. Finally, the DNA content of the new macronucleus is reached by several rounds of replication. Nucleoli reappear and RNA synthesis starts.

Such a cascade of developmental events has been found in *S. lemnae* and *S. mytilus* (Ammermann 1965, 1971), *Euplotes aediculatus* (Ammermann 1971), and *Oxytricha* sp. (Spear and Lauth 1976). There is some evidence that also in *S. muscorum* (Alonso and Pérez-Silva 1965; Pérez-Silva and Alonso 1966), in *S. pustulata* (Oka, pers. commun.), *Euplotes woodruffi*, and *E. eurystomus* (Rao and Ammermann 1970) and perhaps in *Keronopsis rubra* (Ruthmann 1972, own unpubl. data) the development is very similar as described for *S. lemnae*. On the whole, several lines of indirect evidence make it very probable that such a development is characteristic for all hypotrichs.

Thus, macronuclear development in hypotrichous ciliates comprises chromosome elimination in some species and extensive chromatin elimination as well as polytenization in all species investigated so far.

5.3 Molecular Differences Between Macro- and Micronuclear Genomes

5.3.1 DNA Content, Base Composition, and Molecular Weight of DNAs

The gross organization of micronuclear DNA of hypotrichs closely resembles those of other ciliates (Table 1). Micronuclei contain chromosomes and divide by mitosis. The micronuclear DNA content is low compared to corresponding macronuclei. It ranges from 0.07 pg in *Euplotes minuta* micronuclei up to about 12 pg (C-values) in *S. mytilus* (Ammermann and Schlegel 1983; Ammermann, pers. commun., cf. Table 1). One or several micronuclei per cell can be found depending on the species observed. Micronuclear DNA of *Oxytricha* and *Stylonychia* species which has been investigated in some detail has a high molecular weight

Table 1. Differences between macro- and micronuclear genomes of hypotrichous ciliates[a]

Characteristic	Micronucleus	Macronucleus
DNA content in G1[b]	0.14–24 pg	51–1004 pg
Repetitive sequences[b]	27%–55%	Below 2% or absent
Complexity of unique DNA[b]	$\sim 1.5 \times 10^{12}$ daltons	1.7–3.6×10^{10} daltons
Base composition[b, c]	Heterogeneous 36%–50% G + C	Homogeneous 26%–32% (42%)[d] G + C
Satellite DNA fractions[c]	Several present	Absent or reduced
Molecular weight of DNA	Presumably chromosome-sized	About gene-sized DNA molecules 0.3–20 kb[f]
Copy number of individual genes per nucleus	Not determined, presumably two copies per diploid genome for most genes	Varying from 1,000 to more than 10^6 copies per nucleus
Internal eliminated sequences (IES)[e]	Present	Absent
C_4A_4 Repeats	Integrated blocks and terminal repeats	Only terminal repeats at both ends of every gene-sized DNA molecule
Nucleosome repeat length[e]	202 bp	220 bp
Histone proteins	Basic chromatin proteins without similarities with macronuclear or with vertebrate histones	H1, H2A, H2B, H3 and H4 present
Antibodies against Z-DNA[e]	No binding	Strong binding

[a] For references see text.
[b] Minimal vs maximal values reported for hypotrichs.
[c] Determined by buoyant density centrifugation and by melting curve analysis.
[d] The value of 42% G + C reported for *Oxytricha* macronuclear DNA by Lauth et al. (1976) is far outside the range of all other hypotrichs and might be erroneous (?).
[e] So far investigated only in one species of hypotrichs.
[f] Size range within every macronuclear genome.

(greater than about 10–20 kb) and it is assumed to be chromosome-sized (Ammermann et al. 1974; Prescott et al. 1979).

The base composition of micronuclear DNA of those species is heterogeneous, several satellite fractions can be observed in isopycnic density gradient centrifugations (Bostock and Prescott 1972, l.c.). Also thermal denaturation experiments support the finding that micronuclear DNA base composition is heterogeneous (l.c.). Virtually no methylated bases could be detected in micronuclear DNA (Ammermann et al. 1981). Micronuclear DNA base composition differs from that of the corresponding macronuclei (see below).

Micronuclear genomes of *Oxytricha* and *Stylonychia* contain 27% and 55%, respectively, repetitive sequences (l.c., Steinbrück 1976). The complexity of the unique fraction of the micronuclear genomes is in the same range as that found in other eukaryotic genomes (around 2×10^6 kb, corrected for GC differences) (l.c.).

Thus, micronuclear genomes of hypotrichs seem to be "normal" eukaryotic genomes with respect to their DNA characteristics.

Quite in contrast to the apparently normal organization of the micronuclear genomes is the totally different structure of the macronuclear genomes in hypotrichs (see Table 1). The macronuclear DNA content is about 50–100 times higher than the corresponding micronuclear DNA content, ranging from about 50 pg per macronucleus (in G1) in *O. similis* up to about 1,000 pg in *S. mytilus* (Lauth et al. 1976; Steinbrück et al. 1981; Ammermann and Schlegel 1983). The *Oxytricha* species contain, similar to the *Stylonychia* species, two egg-shaped macronuclei per cell. In *Stylonychia* cells, the parts are connected by a thin bridge, and they are, therefore, regarded as forming only one macronucleus. The two parts of the macronucleus of *Stylonychia* as well as the two macronuclei of *Oxytricha* (and the macronuclear parts of other hypotrichous species, too) fuse before every division and replication proceeds synchronously in both parts. Perhaps in *Oxytricha* the connection has been overlooked and there is in fact no difference in the number of macronuclei per cell between *Oxytricha* and *Stylonychia* (see also Sect. 5.1). In *Euplotes* species one band-shaped macronucleus is found in most species.

The base composition of macronuclear DNA of *Stylonychia* and *Oxytricha* is homogeneous, whereas that of micronuclear and anlagen DNA is heterogeneous as indicated by buoyant density centrifugation and by thermal denaturation experiments (Bostock and Prescott 1972; Ammermann et al. 1974; Lauth et al. 1976; Spear and Lauth 1976). The reduction of satellite DNA fractions during anlagen development of *Oxytricha* is accompanied by a shift of the buoyant density of the polytene stage DNA (Spear and Lauth 1976).

Macronuclear DNA of *Oxytricha* and *Stylonychia* species contains about 0.2% ^{6}N-methyl adenine and the GC content ranges from 26% to 37% (Ammermann et al. 1974; Rae and Spear 1978; Steinbrück et al. 1981; Ammermann et al. 1981). Hitherto unexplained discrepancies have been reported to exist between GC values determined by buoyant density centrifugation, melting curve analysis, and direct nucleotide analysis (l.c.). Since the direct quantitative evaluation of chromatographically separated nucleotides or bases gives values of 32%–34% for *Oxytricha* and *Stylonychia* species, the higher values of 42% reported for the same species which as determined by more indirect methods may be incorrect (cf. Ammermann et al. 1974; Lauth et al. 1976).

Renaturation of macronuclear DNA follows second-order kinetics. Therefore, the DNA does not seem to contain significant amounts of repetitive sequences. If present their amount should not exceed 2% (Ammermann et al. 1974; Lauth et al. 1976; Steinbrück et al. 1981). The sequence complexity is low, only 5–11 times larger than that of the prokaryotic *E. coli* genome.

The most striking difference between macro- and micronuclear genomes is found in the molecular weight of their DNAs. Macronuclear DNA molecules range in size from about 0.3 kb to about 20 kb. Most of them have a size of about 3 kb (Prescott et al. 1971, 1973; Rae and Spear 1978; Lawn et al. 1978; Steinbrück et al. 1981). Each of these so-called gene-sized pieces (Prescott et al. 1971) is present on the average 2,000 times in cells of *Oxytricha* (Lauth et al. 1976) and about 15,000 times in cells of *S. lemnae* (Ammermann et al. 1974). It could be shown that different genes occur with differing, gene- and species-specific frequencies in the macronuclei (Steinbrück 1983). These gene- and species-specific frequency patterns lead to species-specific DNA banding patterns after separation of mac-

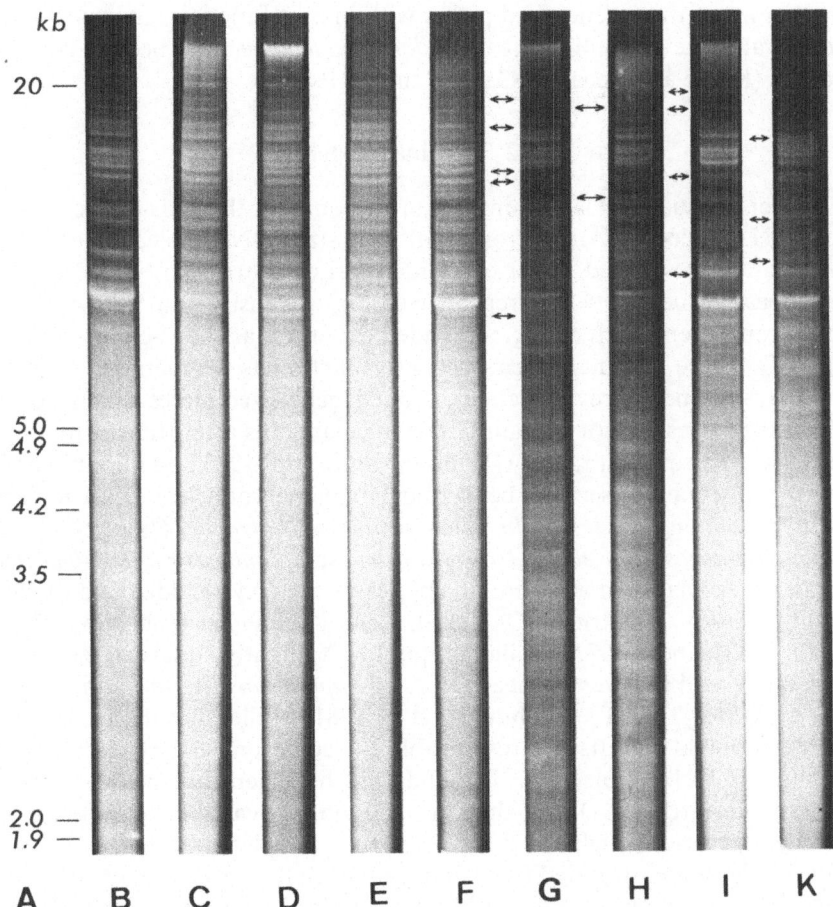

Fig. 7. DNA banding patterns of macronuclear DNA of hypotrichous ciliates. The DNA was separated according to size by electrophoresis in a 1% agarose gel, stained with ethidium bromide, and photographed under UV illumination. *Lane A* gives the fragment size in kb of the marker, λ-DNA, digested with Eco RI and Hind III. *Lanes B–F* show macronuclear DNA of different strains of *Stylonychia lemnae, lane G* of *Euplotes vannus, lane H* of *E. crassus, lane I* of *E. woodruffi, lane K* of *E. eurystomus.* Some species-specific differences in banding patterns are marked by *arrows*

ronuclear DNA of hypotrichs by agarose gel electrophoresis (Lawn et al. 1978; Swanton et al. 1980 b; Steinbrück et al. 1981). Examples of such species-specific macronuclear DNA banding patterns are shown in Fig. 7.

Hybridization experiments with specific gene probes (e.g., rRNA genes, histone genes, actin, and tubulin genes) showed that obviously every single macronuclear DNA molecule contains one gene together with putative regulatory sequences necessary for replication and transcription control and specific telomer sequences at the ends (see Sect. 5.3.3) (Lawn et al. 1978; Elsevier et al. 1978; Lipps and Steinbrück 1978; Spear 1980; Kaine and Spear 1980, 1982; Swanton et al. 1982; Helftenbein 1985). The general structure of macronuclear genes is shown in Fig. 4. Two of these gene-sized pieces with known function, one actin gene of *Oxytricha* and one α-tubulin gene of *Stylonychia lemnae*, have been sequenced in full length (Kaine and Spear 1982; Helftenbein 1985).

5.3.2 Terminal Repeats

Another unusual and surprising feature is found at the ends of every macronuclear DNA molecule. All macronuclear gene-sized DNA molecules of hypotrichous ciliates investigated so far carry identical or nearly identical terminal inverted repeat sequences. These repeats (see Fig. 4) consist of a $5'-C_4A_4C_4A_4C_4$ double-stranded end with a single-stranded extension at the $3'$-ends which is $3'-G_4T_4G_4T_4$. These terminal repeats seem to function as telomers (see Sect. 4.1.5 for comparison and for reviews cited). Cloned, gene-sized pieces which lack such telomere structures cannot be maintained in ciliate cells after transformation for a long time. They are lost rapidly (Wünning and Lipps 1983 and unpubl. results). The terminal sequences seem to be identical in all macronuclear DNA molecules of the hypotrichous species *Stylonychia pustulata*, *S. lemnae*, *Oxytricha* "nova", *O. fallax*, *Paraurostyla weissei*, *Urostyla grandis*, *Keronopsis rubra*, *Pleurotricha spec.*, and presumably most if not all species of the Oxytrichidae and of closely related families of hypotrichs (Oka et al. 1980; Klobutcher et al. 1981; Pluta et al. 1982; Helftenbein 1985; Müller unpubl.). A slightly different terminal sequence was found in macronuclear DNA of *Euplotes aediculatus* and *E. crassus* ending with $3'-G_2T_4G_4T_4$ (Klobutcher et al. 1981; Müller unpubl.).

These terminal repeats are responsible for some unusual characteristics of macronuclear DNA molecules. The ends are hypersensitive against DNase I (Lipps and Erhardt 1981) and they are presumably capable of forming four-stranded aggregates in high salt (Lipps 1980 b). Upon denaturation and quick renaturation the single-stranded DNA molecules can form rings which are held together by a short double-stranded "neck" due to the terminal inverted repeats (Wesley 1975). The single-stranded rings are visible by electron microscopy, but the necks are too small to be detectable under normal DNA spreading and shadowing conditions.

A further consequence of the peculiar terminal repeat structures might be seen in the reactivity of macronuclei with antibodies directed against Z-DNA (Lipps et al. 1983). Antibodies specific for left-handed Z-DNA react strongly with macronuclei of *S. lemnae*, whereas micronuclei and anlagen with giant chromosomes do not react. The reactivity of the macronuclear DNA seems to be dependent on both the presence of the terminal repeats and the integrity of the gene-sized DNA molecules as indicated by Bal 31 and short DNase I digestion experiments, respectively. Whether the reaction with anti-Z-DNA antibodies really indicates that in hypotrichous macronuclei in vivo DNA in Z-configuration exists, as the authors assume, must await further supporting evidence. It cannot be excluded that the differences found in the reactivity with anti-Z-antibodies are due to fixation artifacts (Hill and Stollar 1983).

The organization of the C_4A_4 repeats in the corresponding regions of the micronuclear genomes are treated in the following section.

5.3.3 Comparison of Individual Sequences

A few macro- and micronuclear DNA sequences have been compared by using molecular cloning techniques. No DNA sequences with known protein coding functions from both macro- and micronuclear genomes have been sequenced so far. Detailed sequence comparisons should help to unravel the molecular mechanisms which cause the dramatic genome reorganizations during nuclear differentiation in hypotrichous ciliates.

Since all macronuclear DNA molecules in hypotrichs carry terminal inverted repeats (see Sect. 5.3.2) the question arose whether these repeats flank the integrated micronuclear counterparts of the macronuclear genes. The attractive idea behind this question was that the repeats may serve as recognition signals for excising the macronuclear genes from the giant chromosomes during macronuclear development (Lawn et al. 1978; Dawson and Herrick 1982).

Initially, hybridization experiments with C_4A_4 probes did not give conclusive results. Boswell et al. (1982) reported that C_4A_4 repeats are not present at the junctions of macronuclear genes with flanking sequences in the micronuclear genome of O. "nova". But Dawson and Herrick (1982) do not rule out this possibility with certainty due to their hybridization experiments with O. fallax micronuclear DNA.

Furthermore, detailed sequence comparisons of macronuclear genes with their micronuclear counterparts revealed that no C_4A_4 repeat blocks are detectable at the corresponding regions of the micronuclear genomes (Oka and Honjo 1983; Klobutcher et al. 1984). Since in the study of Oka and Honjo (l.c.) the compared sequences differ by a few percent of the bases, the authors might not have taken homologous sequences from the genomic libraries. The micronuclear gene they used might be a nonprocessed "pseudogene". But that problem has been taken into account by Klobutcher et al. (l.c.) and their results are fully convincing.

Long C_4A_4 homologous repeat blocks were found in other regions of the micronuclear genome of O. fallax which are likely to be telomeres of micronuclear chromosomes (Dawson and Herrick 1984) due to Bal 31 sensitivity. In addition, a few hundred internal C_4A_4 repeat blocks containing only 20–40 bp were detected. Only a few of them seem to be contiguous with macronuclear genes.

Thus, there now seems to be agreement that macronuclear DNA termini are generated either by some recombination during the elimination process or by a step-by-step addition after excision.

Another interesting feature of the organization of macronuclear genes in the micronuclear genome was found by Boswell et al. (1983). Macronuclear genes are not dispersed randomly in the micronuclear genome of O. "nova" and separated from each other by long stretches of "spacer" sequences. Instead, macronuclear genes are clustered in groups of two to five genes in micronuclear DNA of this species.

Sequence comparisons of cloned macronuclear genes with the corresponding micronuclear counterparts gave another interesting aspect of nuclear differentiation in O. fallax. A family of macronuclear genes was found whose members share a common sequence block (Cartinhour and Herrick 1984a). Sequence analysis of the corresponding micronuclear DNA fragments revealed that the members of the macronuclear gene family may be generated by alternate processing during macronuclear development (Cartinhour and Herrick 1984 b).

Evidence for the most exciting molecular reorganization process during macronuclear development was presented by Klobutcher et al. (1984). Two cloned macronuclear genes were collected from a macronuclear genomic library. They differ in about 3% of their sequences and are, therefore, assumed to be alleles (the macronucleus develops from a presumably diploid micronucleus) or members of a small gene family. Their sequences were compared with the putative micronuclear precursor sequences. In contrast to expectations, the macronuclear genes were not found to be colinear with their micronuclear counterparts. Three short internal sequence blocks must be removed during development by a presumptive DNA splicing process. These internal blocks are called internal eliminated sequences (IES) to distinguish them from introns which are excised from RNA molecules. The IES show features which are characteristic of transposable elements found in other organisms (e.g., Calos and Miller 1980). Short, direct repeats with adjacent inverted repeats are found at the ends of each of the IESs. These features may play a role in the excision process. The putative functional role of the IES will be discussed later (Sect. 8.2). The detection of the IES elements provides an entire novel aspect to the complex process of nuclear differentiation in ciliates.

The junctions of the macronuclear genes with the flanking micronuclear DNA sequences exhibit some interesting features, too. It has been mentioned above that C_4A_4 repeats are absent at these locations. The flanking micronuclear sequences contain a large number of inverted repeat sequences which might form specific secondary structures in vivo, necessary for the excision process. In addition, an imperfect, directly repeated sequence is found at each junction. A similar structure was found at the junctions investigated by Oka and Honjo (l.c.). Whether these structures are necessary prerequisites for the excision process must await new experiments which may elucidate the functions of these junction sequences.

5.3.4 Chromatin Structure and Histones

Brief digestion of macro- and micronuclear chromatin of *S. lemnae* with micrococcus nuclease revealed that differences between micro- and macronuclei exist in the nucleosome repeat length (Lipps and Morris 1977). Macronuclear chromatin has a repeat length of about 220 bp and micronuclear chromatin one of about 202 bp. Since a longer digestion gives an identical nucleosome core length of about 140 bp for both nuclei, one can assume that the difference lies in the linker DNA between the nucleosome cores. The difference in nucleosome repeat length between transcriptionally active macronuclear and inactive micronuclear chromatin does not agree with observations of changes in nucleosome repeat length in some other eukaryotes. In chicken and sea urchin, for instance, transcriptionally inactive tissues have chromatin with longer repeats than transcriptionally active tissues (l.c.).

Differing histone fractions were found in macro- and micronuclei of *Stylonychia* and *Oxytricha* (Lipps et al. 1974; Lipps and Hantke 1975; Caplan 1975, 1977). Macronuclei seem to contain histones which resemble the known eukaryotic histone fractions H2A, H2B, H3, and H4. But whereas H1 of *Stylonychia* also resembles H1 of other eukaryotes, H1 of *Oxytricha* has been reported to have an unusually high electrophoretic mobility in acid acrylamide-urea gels. This discrepancy is as yet unexplained. It might be caused by misidentification of histone fractions in one of the reports.

Micronuclear chromatin contains basic proteins which can not be homologized with known histone fractions of other organisms (cf. Sect. 4.1.3). In *Oxytricha* these basic proteins are very lysine-rich and resemble perhaps basic proteins found in sperm of some animals (Caplan 1977).

Between macronuclear and anlagen chromatin of *Stylonychia* only slight quantitative differences in some histone fractions seem to occur. But one should keep in mind that anlagen preparations are in most cases heavily contaminated with macronuclear material. No controls for assaying such contaminations were included in the experiments reported (l.c.).

Since histones may play an important role in the functional differences of macronuclei, micronuclei, and anlagen, more detailed analyses are highly desirable for the hypotrichs.

5.4 Putative Somatic Functions of the Germ Line Nuclei of Hypotrichs

There are also weak indications for hypotrichs that micronuclei may have essential vegetative functions in addition to their well-known generative functions during sexual reproduction (see Chap. 2; cf. Sects. 4.1.6 and 4.3.3). Ammermann (1970) presented evidence for ^{3}H-uridine incorporation into micronuclei of *Stylonychia*. It seems doubtful whether the autoradiographic label over isolated micronuclei is really due to RNA synthesis in vivo. It might be the result of cytoplasmatic contamination of the isolated nuclei or newly synthesized RNA might originate from macronuclei and might have migrated into the micronuclei. As the author admits, neither possibility can be ruled out.

Several studies clearly indicate that removal of micronuclei causes severe deficiencies in vegetative cells of hypotrichs (Ammermann 1970; Mikami et al. 1985). This is comparable to the situation found in other groups of ciliates (see, e.g., Ng and Newman 1984). Cells whose micronuclei have been removed by irradiation or by microsurgery often degenerate after one or a few divisions. They become unable to move, to feed, and finally die. Only a very low number of cells recover and survive without micronuclei. There are indications that normal stomatogenesis which occurs during every cell division and after conjugation is impossible or heavily affected in most amicronucleated cells.

The question remains open why some amicronucleate cells nevertheless are viable. Perhaps those viable amicronucleate cells may have conserved some essential micronuclear components within their cells and, therefore, the macronuclei may be assumed to have taken over in these rare cases vegetative micronuclear functions, e.g., in stomatogenesis. Thus, the hypothesis should be tested by suitable experiments whether in viable amicronucleate clones the macronuclei have been transformed with micronucleus-specific sequences which might originate from destroyed micronuclei.

Finally, it must be questioned whether a putative vegetative function of the micronuclei must be related in any case to transcriptional activity. One may imagine that the possible vegetative functions of micronuclei might also be due to their binding capacity for regulatory molecules with a probable dosis effect.

5.5 Concluding Remarks

The hypotrichous ciliates exhibit the most dramatic molecular reorganization during nuclear differentiation. In all species investigated, giant chromosomes are found, extensive chromatin elimination takes place, and differential amplification is involved in macronuclear development. In addition, in some species the elimination of whole chromosomes contributes to the differences between germ line and soma genomes. Also chromatin structure and histone composition changes during that process. The evolution of nuclear differentiation in ciliates has reached the top position with the Hypotrichida.

The most prominent molecular differences between macro- and micronuclear genomes of hypotrichous ciliates are summarized in Table 1.

6 Indications for Molecular Differences Between Macro- and Micronuclear Genomes in Other Groups of Ciliates

Practically no relevant data on molecular events occurring during nuclear differentiation are available for species belonging to other orders except the Hymenostomatida (see Chap. 4) and the Hypotrichida (see Chap. 5). Nevertheless, a few facts on nuclear development in other groups of ciliates shall be mentioned here. Despite being very fragmentary, the data may allow some conclusions on underlying molecular processes. They seem to be meaningful for a discussion of nuclear differentiation in ciliates from a general point of view (see Chaps. 7–9).

6.1 Bursariomorphida

Bursaria truncatella is a large ciliate nearly 1 mm in length with tiny micronuclei and very large, DNA-rich macronuclei. It was believed for a long time to belong to the Heterotrichida (see Sect. 6.4). But recently it was recognized as a member of a separate order, Bursariomorphida (Fernandez-Galiano 1979), of the class Kinetofragminophora. The class Kinetofragminophora comprises among others, the orders Karyorelictida (Chap. 3), Cyrtophorida (Sect. 6.2), and Suctorida (Sect. 6.3).

If the assumption is correct that large ciliate cells, which require a high amount of actively transcribable genetic material, eliminate larger amounts of the germ line genome during nuclear differentiation than species with smaller cells, *B. truncatella* should be an excellent example to test that hypothesis (see Chap. 8

for discussion). Macronuclei of *B. truncatella* contain about 5,000 times the haploid DNA content (C-value) of the micronuclei (Ruthmann 1964). Comparisons of the DNA complexities of both kinds of nuclei are still lacking. Only a preliminary biochemical characterization of macronuclear DNA was done (Borchsenius and Sergejeva 1979). An estimation of the molecular weight of macronuclear DNA by centrifugation in alkaline sucrose gradients revealed two molecular weight fractions: a larger one with molecules of about 10×10^6 daltons and a smaller one with molecules of about 100×10^6 daltons. Macronuclear DNA of *Tetrahymena* under comparable conditions had a molecular weight of 400×10^6 daltons (cf. Sect. 4.1.2) Whether this relatively low molecular weight of *B. truncatella* macronuclear DNA can be taken as an indication for a fragmentation during macronuclear development cannot be decided without further experiments. The past observation made by Poljansky (1933) together with more recent autoradiographic investigations (Poljansky and Sergejeva 1981) seem to be worth mentioning in this connection. During the anlagen development of *B. truncatella*, an "achromatic" stage (cf. Sects. 6.2 and 6.3) is observed and a large heterochromatic chromatin block is formed which later disappears. At the beginning of the anlagen development chromosomes become visible which presumably become oligotenic and disappear later. Two phases of DNA synthesis are separated by the "achromatic" stage. This pattern of DNA synthesis during macronuclear development resembles that found in hypotrichs (cf. Fig. 5).

Combining the fragmentary molecular data concerning macronuclear development of *B. truncatella*, one may conclude that indications for the occurrence of chromosome fragmentation and chromatin elimination are also given in this species. Hybridization experiments with single gene probes onto electrophoretically separated macronuclear DNA and sequence comparisons of cloned macro- and micronuclear genes are necessary to clarify this process.

6.2 Cyrtophorida

The genus *Chilodonella* belongs to the order Cyrtophorida of the class Kinetofragminophora (Corliss 1979). No close relationship seems to exist either to the Hymenostomatida (Chap. 4) or to the Hypotrichida (Chap. 5). The peculiarities of the macronuclear development of *Chilodonella cucullulus* studied by Radzikowski (1973, 1976, 1979) are, therefore, of especial interest. He detected giant chromosomes which form during anlagen development in *C. cucullulus*. DNA replication is asynchronous along the giant chromosomes. Those segments of the giant chromosomes where intensive DNA replication can be observed, closely resemble DNA puffs as they are found in Sciaridae (Pavan and DaCunha 1969; Glover et al. 1982; Amabis 1983). After fragmentation only those fragments which contain the structures resembling DNA puffs are maintained and give rise to the final macronucleus genome. The rest of the giant chromosomes fuses into a dense chromatin body (cf. Sect. 6.1) which is eliminated during the first postconjugational division. The chromatin elimination comprises about 35% of the anlagen chromatin. Thus, in *C. cucullulus* a very similar cascade of developmental events is found during nuclear differentiation as in hypotrichs. Macronuclear development comprises formation of giant chromosomes, chromatin elimination, and differential amplification processes (cf. Sect. 5.2 and Fig. 5).

Giant chromosomes have also been found in the anlagen development of the related species *C. steini* (Radzikowski and Golembiewska 1977).

The macronuclear development of the related species *C. uncinata*, however, does not show such a complex pattern of events (Pyne et al. 1974; Pyne 1978). No giant chromosomes are visible, and no DNA elimination seems to occur. But the occurrence of small chromatin granula in the macronuclear anlage which are not connected to each other as well as the appearance of an "achromatic" phase (cf. Sects. 6.1 and 6.2) might also give an indication of chromatin fragmentation and elimination in this species. Obviously, the interesting anlagen development of the *Chilodonella* species deserves reinvestigation by biochemical techniques.

6.3 Suctorida

The Suctorida belong to the class Kinetofragminophora, as the two orders mentioned above (Corliss 1979), but their phylogenetic relationship is controversial. Suctors are very aberrant ciliates. They are sessile and feed with tentacles rather than with a cytostome. Cilia are absent in adult animals, only juvenile forms have cilia and are mobile. However, also in species of this aberrant order, a weak indication for putative molecular changes during nuclear differentiation can be found. Grell (1949, 1953) detected that early in the anlagen development of *Ephelota gemmipara* bundles of chromosomes are formed by successive rounds of replication. These boundles of chromosomes partially fuse, forming a so-called Binnenkörper or karyosome. This karyosome is finally resorbed. Later an "achromatic" stage follows during which the anlagen are less stainable with Feulgen stain. Finally, DNA synthesis starts again (cf. Sects. 6.1, 6.2, 5.2). Whether these events can be interpreted as oligotenization, DNA elimination, and subsequent differential amplification remains to be tested by molecular methods.

6.4 Heterotrichida

The order Heterotrichida belongs to the same class Polyhymenophora as the Hypotrichida (Chap. 5). Due to many lines of evidence, these two orders can be assumed to be more closely related to each other than to any other group of ciliates mentioned in this article. Some fragmentary data regarding nuclear differentiation seem to be worth mentioning since they are similar to phenomena observed in the related order Hypotrichida.

During the macronuclear development of the heterotrich *Nyctotherus cordiformis* giant chromosomes have been observed (Grassé 1952; Golikova 1965). Their appearance is similar to that of the giant chromosomes of hypotrichs and of *Chilodonella*. But no biochemical investigations of nuclear differentiation have been reported so far.

It is astonishing that macronuclear development of some of the largest ciliates, e.g., the heterotrichs *Stentor* or *Spirostomum*, has not been investigated by molecular methods. These heterotrichs contain very small micronuclei, but very large, DNA-rich macronuclei. The reasons for the lack of this type of investigation are that conjugation is very rarely found in these species and that considerable technical problems exist for isolating the necessary quantities of pure micronuclei. A coarse comparison of the sequence complexities of both kinds of nuclei of one of these species might give a first indication toward the occurrence or absence of sequence elimination during nuclear differentiation. Unfortunately, sequence complexities have not been measured by reassociation kinetics in any of the heterotrichs of either nuclei. Only the macronuclear DNA complexity of *Stentor coeruleus* has been determined by re-association experiments (Pelvat and DeHaller 1976). It was found to be 6×10^{10} daltons (ca. $21 \times E.\ coli$). This value lies in the range known for macronuclear DNA complexities of other ciliates (e.g., Soldo et al. 1981; Steinbrück et al. 1981). For the micronuclear genome only the DNA content measured by microspectrophotometry is known. The C-value of micronuclei of *S. coeruleus* is 0.16 pg (Ammermann, unpubl.). A comparison of these two values (0.16 pg $= 9.7 \times 10^{10}$ daltons and 6×10^{10} daltons) implies that the analytical complexity of the micronuclear genome seems to be larger by a factor of about 1.6 than the sequence complexity of the macronuclear genome. Of course, both values are not exactly comparable to each other, since the amount of repetitive sequences of the micronuclear genome is unknown. The macronuclear genome contains about 15% repetitive sequences (l.c.). Measurements of micronuclear DNA sequence complexities of *Stentor* and of other heterotrichs would be highly desirable. Nevertheless, such a coarse comparison can be assumed to give an underestimate if the difference is not the result of experimental errors. The actual difference might even be larger. If this is true, the difference can be explained in two ways. Either *Stentor* eliminates chromosomes and/or chromatin during macronuclear development, which has not been reported so far, or differential amplification takes place. No evidence has been found in the investigation of macronuclear DNA of *S. coeruleus* (l.c.) for the occurrence of low molecular weight DNA which would be a clear indication for fragmentation. But detailed comparisons of macro- and micronuclear DNA sequences with the help of molecular cloning methods seem to be very promising.

7 Origin and Evolutionary Trends of Nuclear Differentiation

Evolutionary trends are difficult to follow in ciliates, since the phylogenetic relationships between ciliates are far from being clear. Despite this disappointing situation, a discussion or at least speculation on some evolutionary aspects of nuclear development seems to be indicated. On the basis of the information presented in the previous chapters, the following questions shall be discussed.

1) Has nuclear dualism originated within the class of ciliates or among nonciliate ancestors?
2) Which of the ciliates existing today show the most primitive form and which the most advanced form of nuclear dualism?

3) Did the complex events of nuclear differentiation, e.g., formation of giant chromosomes, extensive DNA elimination, and changes in DNA composition originate several times independently in ciliates?

7.1 Origin of Nuclear Dualism

Since nuclear dualism is found not only in ciliates, but also in some foraminiferans, the first question can be treated adequately only if the nuclear differentiation of foraminiferans is taken into account. Such a comparision can perhaps elucidate whether nuclear dualism as it is found in ciliates is a unique peculiarity of that group or whether it has been evolved earlier in eukaryotic evolution. First the phylogenetic relationships of foraminiferans must be considered briefly. According to "A Newly Revised Classification of the Protozoa" (Levine et al. 1980) the Foraminiferida are an order of the superclass Rhizopoda, the amoebae, which form together with the flagellated protozoans the phylum Sarcomastigophora. The relationship of that phylum to the phylum Ciliophora, comprising all ciliates, is absolutely unclear. Also the question whether the Foraminiferida form a monophyletic group is obviously unresolved (Berthold, pers. commun.).

Because no established phylogenetic relationship elucidates the origin of nuclear dualism, perhaps a comparison of the nuclear dualism in foraminiferans with that in ciliates might indicate its origins.

Nuclear dualism has been found only in some polythalamous foraminiferan species (for reviews, see Grell 1979; Raikov 1982). In those species small generative and large somatic nuclei are found within one cell in various stages of their complex life cycles. Function and fate of the two kinds of nuclei resemble the situation found in ciliates (cf. Chap. 2). The division products of a zygote nucleus differentiate into two types of nuclei: one or several somatic nuclei and several generative nuclei. The somatic nuclei have in some, but not in all, species a DNA content which can be up to 15 times larger than that of the corresponding generative nuclei (Raikov 1982). In the species *Rotaliella heterocaryotica* the somatic nucleus contains several times more RNA than the generative nucleus (Zech 1964). The somatic nuclei contain nucleoli and are active in transcription. During the meiotic divisions of the generative nuclei, the somatic nuclei degenerate. On the other hand, generative nuclei are relatively small and compact. In most, but not in all, species they seem to contain nucleoli. Their RNA content is low and remains unchanged [investigated only in one species by Zech (1964)]. Unfortunately, detailed molecular analyses of nuclear differentiation in foraminiferans are lacking.

The resemblance of nuclear dualism in ciliates and in some foraminiferan species can be explained by several hypotheses:

1) Nuclear dualism as well as the developmental events leading to it have evolved independently of each other in both groups. The resemblance is due to similar functional constraint. In other words, it is a pure convergency. This hypothesis seems to be most favored today, but its support is derived mainly from lack of evidence.

2) Foraminiferans or at least some of them are phylogenetically more closely related to ciliates than generally assumed by systematicists. The resemblance is due to a common origin of nuclear dualism in the common ancestor of both groups. If the foraminiferans are a polyphyletic group, only a subgroup must be considered. All ciliates would have conserved the characteristics inherited from the ancestor, most foraminiferans have lost them. In this case nuclear dualism in both groups would be truly homologous.
Confirming evidence supporting this hypothesis seems to be lacking so far.

3) Nuclear differentiation is a basic feature of eukaryotic evolution. It evolved before eukaryotes were separated into the major systematic groups. In this case the resemblance in nuclear dualism could be due to parallel evolution (see Hennig 1966 for discussion). Parallel evolution leads to character correspondences which have been acquired independently, but in parallel in related groups. The reason for the parallel evolution is the common genetic background inherited from the common ancestor. This distinguishes parallel evolution from normal convergent evolution.

According to the third hypothesis, a common ancestor of ciliates, foraminiferans, and perhaps of multicellular eukaroytes also must already have evolved molecular mechanisms leading to nuclear differentiation. In ciliates and some foraminiferans this common heritage gave rise to nuclear dualism, in multicellular organisms it led to nuclear differentiation in separate cells.

In this context it should be mentioned that some authors assumed that ciliates are closely related to metazoans (e.g., Hadži 1963; Hanson 1963; Steinböck 1963). Arguments in favor of this assumption are, among others, the diploid life cycle and the gametic meiosis. Supporting evidence for this "ciliate theory" of the origin of metazoa has been presented recently by 5S-RNA sequence comparisons (Ohama et al. 1984). Perhaps modifying this hypothesis in assuming that ciliates and metazoans might be sister groups, would make the idea more meaningful and could give rise to a more serious discussion [for arguments against this idea see, for instance, Möhn (1984)].

Only more molecular investigations on the mechanisms of nuclear differentiation in foraminiferans as well as in so-called primitive ciliates might contribute essential evidence for confirmation of one of the three alternatives with regards to the roots of nuclear differentiation in eukaryotes.

7.2 Trends in the Evolution of Nuclear Differentiation in Ciliates

As to which group of the ciliates shows the most ancient and which the most advanced form of nuclear dualism is also suffering from fragmentary experimental data as well as from the unresolved phylogenetic relationships among ciliates. Nevertheless, some aspects of this problem are noteworthy of discussion.

The problem of the most primitive form of nuclear dualism among ciliates is closely connected with the answer to the question as to which of the existing ciliates form the most ancient group, i.e., the group which has retained the majority of original features. It seems to be accepted by most authors, now, that the Karyorelictida are the most primitive group of ciliates. For a detailed discussion of that problem see Corliss (1979) and Corliss and Hartwig (1977). But even if that view is accepted, one should keep in mind that the karyorelictids which live today have participated in evolution as long as any other species which now exist. Therefore, they may have evolved very advanced specializations, too, despite having conserved many ancient features. Such a mosaic evolution is found frequently.

For instance, adaptations to their unique environment or the peculiar nuclear complexes are considered as highly specialized characters (see Chap. 3). It is, therefore, not obviously clear whether their form of nuclear dualism having DNA-poor nearly diploid macronuclei is really the most ancient form among ciliates. But several arguments seem to be in favor of this view.

As a first step toward nuclear dualism, the evolution of homokaryotic species can be postulated. Homokaryotic ciliates, i.e., ciliates with two or more identical nuclei within one cell, are unknown (see Chap. 2). The next step in the evolution of nuclear dualism must be species with two sorts of nuclei which are already different in function, but have remained diploid or nearly so. Exactly that theoretically postulated nuclear situation is found in most karyorelictids. The peculiar characteristics of karyorelictid macronuclei, e.g., the inability to divide must be viewed as specializations.

Another quite convincing argument is put forward by Corliss in his critique of Fauré-Fremiet's opinion (Corliss 1979; Corliss and Hartwig 1977; Fauré-Fremiet 1954). Fauré-Fremiet assumes that the adaptation of the karyorelictids to their biotope, the mesopsammal, has required a reduction in macronuclear size and its DNA content accordingly. However, it seems highly improbable according to Corliss that the DNA-poor macronuclei of the karyorelictids are specialized adaptations to their environment and that they evolved from DNA-rich macronuclei. On one hand, ciliates belonging to other systematic groups live in the same environment with DNA-rich macronuclei. On the other hand, no convincing example of animal evolution is known that species with polyploid or DNA-rich genomes have returned to the original diploid state or to very DNA-poor genomes. According to current knowledge, the DNA-rich genomes are the more specialized genomes in a group of related species. As a consequence of that one-way direction in evolution, species with DNA-rich genomes are on a dead end pathway. Obviously, ciliates have circumvented that problem by their nuclear dualism.

Considering all the arguments discussed above, the opinion seems to be justified that the form of nuclear dualism as it is found in karyorelictids is in fact the most ancient form and closest to the origin of nuclear dualism. The evolution from the assumed homokaryotic to the first heterokaryotic species must have taken place before the karyorelictids originated, and perhaps even before ciliates evolved.

The question of the most advanced form of nuclear dualism can be more easily answered. If evolution proceeded from DNA-poor diploid macronuclei toward highly amplified genomes, undoubtedly the hypotrichous species show the most advanced type of nuclear dualism found in ciliates thus far. In species of the families Oxytrichidae, Euplotidae, and Urostylidae the greater parts of the micronuclear genomes are eliminated during macronuclear development and the remaining fragments of the genomes are amplified up to more than 10,000 times in some species (see Chap. 5). Further progress of this evolutionary trend in that direction in ciliates seems to be impossible. Therefore, nuclear dualism as it is found in some hypotrichs seems to be an endpoint of that evolutionary pathway.

The question remains open whether the hypotrichous ciliates have made a sidestep in ciliate evolution. Until recently, most ciliatologists seemed to favor the opinion that the vast majority of the ciliate species possess a truly polyploid macronucleus (e.g., Grell 1973; Raikov 1976; Corliss 1979) whereby the hypotrichs are an exception. This view implies that the micronuclear genome survives nuclear differentiation in most species unchanged and that, therefore, macronuclei contain more or less unchanged multiples of the micronuclear genomes. I do not think that the molecular data gathered during the last few years support that view. It was shown in the foregoing chapters that molecular changes occurring during macronuclear development could be detected in nearly every species which was investigated more carefully by molecular and biochemical techniques. The only exception thus far among the better investigated species might be some species of the *Paramecium aurelia* species complex (see Sect. 4.3). With exception of rDNA amplification and one report about small-sized macronuclear DNA (McTavish and Sommerville 1980), no changes occurring during macronuclear development have been reported for that group so far.

7.3 Did the Complex Features of Nuclear Differentiation Arise Several Times Independently Among Ciliates?

In several groups of ciliates nuclear differentiation is tightly connected with complex intranuclear events, e.g., formation of giant chromosomes, DNA elimination, differential amplification, sequence rearrangements, and appearance of replication bands. Systematicists seem to agree that these groups of ciliates are not phylogenetically closely related (e.g., Grell 1973; Corliss 1979). Giant chromosomes are found in the orders Cyrtophorida (Sect. 6.2), Heterotrichida (Sect. 6.4), and Hypotrichida (Sect. 5.2) which belong to two different classes. DNA elimination has been reported to occur in species of the orders Cyrtophorida (Sect. 6.2), Hymenostomatida (Sect. 4.1.5), and Hypotrichida (Sect. 5.2) which belong to three different classes. It occurs perhaps in other orders, too (e.g., Chap. 3, Sects. 6.1 and 6.3). Differential amplification seems to occur in Karyorelictida (Chap. 3), Cyrtophorida (Sect. 6.2), Suctorida (?) (Sect. 6.3), Hymenostomatida (Sects. 4.1.5, 4.2, 4.3), Heterotrichida (?) (Sect. 6.4), and Hypotrichida (Sect. 5.2). Replication bands are found in Cyrtophorida, Chonotrichida, and Hypotrichida. Of course, the phylogenetic relationships among ciliates might have been misinterpreted and may not reflect the actual relationship. However, if systematicists are not totally wrong, another interpretation of the distribution of developmental characteristics must be found. The simplest interpretation would be convergency, i.e., totally independent evolution in different species groups due to similar functional requirements. Convergent evolution cannot be ruled out. However, this assumption seems to be somewhat problematic since the basic molecular mechanisms are very complex and a large amount of independent, but parallel, mutational changes would be required.

Another assumption appears more likely. The basic molecular mechanisms might be a fundamental property of all ciliates or perhaps of most eukaryotes (see Sect. 7.1). In that case, primitive ciliates or their ancestors must already have evolved those mechanisms. Only in some species does this molecular equipment lead to detectable DNA elimination, differential amplification processes, or the formation of giant chromosomes, among others. But in all ciliates these capacities would be the basic condition of nuclear differentiation. Only the extent to which they produce differences between macro- and micronuclear genomes differs from species to species. Whether this hypothesis can be confirmed remains to be seen after further investigations. Macro- and micronuclear genomes of a greater number of species belonging to different taxa should be compared by biochemical methods, e.g., restriction mapping and sequence comparisons of macro- and corresponding micronuclear genes.

The main points discussed in this chapter may be summarized as follows. Nuclear dualism is characteristic of all ciliates. Evolution of nuclear dualism in ciliates seems to have proceeded from species with nearly identical nuclei toward species with extremely different nuclei. DNA elimination, differential amplification, and sequence rearrangements seem to belong to the basic features of nuclear differentiation, in most, if not all species. Some observations suggest that these mechanisms might be an ancient heritage of the majority of eukaryotes which is found in the most prominent form in some ciliates.

8 Putative Functions of Molecular Reorganization Processes During Macronuclear Development

The experimental results presented in the foregoing chapters allow one clear conclusion. The dogma of the constancy of DNA composition throughout the life cycle of an organism does not hold true for the ciliates. Nuclear differentiation produces in many ciliate species more or less prominent differences with regard to many characteristics of the genetic material of both kinds of nuclei. If in a flourishing group of animals such a central dogma – the constancy of the genetic material – is broken, the investigation of the reasons for this occurrence would be one of the most exciting aspects of the whole spectrum of the existing problems. From an evolutionary point of view the reasons must lie in a positive selection of advantageous functions which are fulfilled by the molecular reorganization processes. Experimental access to the putative functions of molecular reorganization processes is very difficult. To date, investigations of the mechanisms of nuclear differentiation in ciliates are still in a descriptive phase. However, some putative functions become faintly visible. A discussion of putative functions might be useful in order to stimulate and encourage the necessary experiments. Indispensable techniques are now available due to recent progress in molecular cloning and modern biochemical analytical techniques.

8.1 The C-Value Paradox

8.1.1 Did Ciliates Receive a Specific Solution of the Problems Raised by the Micronuclear C-Value Paradox?

A key point in the search of putative functions of molecular reorganization during nuclear development seem to be the widely varying DNA contents of micro- and macronuclei (cf. Table 2). Among ciliates many examples of the so-called C-value paradox can be found. The C-value paradox means that in many groups of organisms C-values (= amount of DNA per haploid genome) are found which differ by orders of magnitude between species which exhibit the same general level of morphological and functional complexity (for a recent review see Gall 1981).

C-values found in micronuclear genomes of hypotrichous species can be taken as an example of the C-value paradox in ciliates. Micronuclei of *Euplotes minuta* have a C-value of 0.07 pg (Ammermann, unpubl.), whereas the C-value of micronuclei of *Stylonychia lemnae* is about 130 times larger (9.2 pg; Ammermann and Schlegel 1983). It seems very unlikely that despite a considerable difference in cell size both species have so differing genetic requirements. The C-value paradox is as yet unresolved. Some putative explanations are discussed later (Sects. 8.1.2, 8.1.3, 8.1.4). But perhaps one function of molecular reorganization is closely correlated with the C-value paradox. DNA elimination which is found in macronuclear development of several groups of ciliates (see Chap. 3, Sects. 4.1.5, 4.2, 5.3, 6.2) might provide a solution of the problems arising with large micronuclear C-values. It was mentioned before that macronuclei which descend by nuclear differentiation from micronuclei are in most species rich in DNA. It is reasonable to assume that the differences in micronuclear C-values are not due to different numbers of transcribable, protein-coding genes. Therefore, a large number of DNA sequences which are dispensable for macronuclear functions together with the necessary sequences would have to be amplified in species with high micronuclear C-values if no elimination or underreplication occurs. Species which eliminate DNA sequences might have resolved that problem by eliminating the dispensable and amplifying only the essential sequences.

It might be considered as supporting evidence that macronuclear DNA sequence complexities are very similar despite very different DNA contents. The complexities of macronuclear genomes do not differ very much and approx. 3×10^{10} daltons are found for many of the species investigated (cf. Table 2). Unfortunately, analytical data are insufficient in elucidating whether species with low micronuclear C-values eliminate fewer or no DNA sequences and species with larger micronuclear C-values eliminate larger amounts. An investigation of the nuclear differentiation of *Euplotes minuta* would be extremely interesting under this aspect. *E. minuta* with its low C-value (see above) might have a micronuclear genome complexity which is in the range found for macronuclear genomes of hypotrichs. According to the hypothesis presented, only a low amount or no elimination at all should be found in this species in contrast to the *Euplotes*, *Oxytricha*, and *Stylonychia* species with higher micronuclear C-values (see Sect. 5.2).

The suggestion discussed above immediately evokes several questions: Why do ciliate macronuclei contain such large amounts of DNA in many species? What about the ciliate species which do not seem to eliminate larger amounts of DNA? Did they evolve another strategy to circumvent the economical problems raised by the C-value paradox or are their micronuclear C-values lower?

8.1.2 The Skeletal DNA Hypothesis

The macronuclear DNA content in ciliates ranges from near diploid amounts in karyorelictids (see Chap. 3) to enormous amounts in some hypotrichs and heterotrichs (see Table 2) and also a few species in other groups. Whereas the micronuclear DNA content is not correlated with cell size, a positive correlation between macronuclear DNA content and cell size exists in hypotrichous ciliates (Ammermann and Münz 1982) and presumably in other ciliates, too (Raikov 1982). Larger ciliates always have higher macronuclear DNA contents than smaller ones.

The correlation between DNA content and cell size is reminiscent of the skeletal DNA hypothesis of Cavalier-Smith (1978, 1982), which claims to provide an

Table 2. Nuclear DNA contents and genome complexities

	Micronuclear DNA content in pg (2C)	Macronuclear DNA content in pg (in G1)	Macronuclear genome complexity relative to *E. coli*[a] $(=2.8 \times 10^9$ daltons)[b]
Tetrahymena thermophila	0.46[c, +]	10[c, +]	28.5 × [c, +]
Paramecium primaurelia	0.63[d]	70[e]–274[d]	11[e]–19[f] ×
P. bursaria	7.5[g]	90[g]–277[h]	3.9[h]–33.6[g] ×
Stylonychia lemnae	18.4[i]	788[i, j]	11.1 × [h]
S. mytilus	24.8[i]	1004[i, j]	_[k]
Oxytricha "nova"	1.3[l]	115[j, l]	13 × [l]
O. similis	_[k]	51[h, j]	6.1 × [h]
O. fallax	1.1[m]	132[j, m]	_[k]
Paraurostyla weissei	8.0[n]	348[h, j, m]	5.0 × [h]
Euplotes aediculatus	2.0[n]	380[h, n]	10 × [h]
E. minuta	0.14[o]	85[o]	_[k]
Stentor coeruleus	0.32[o]	2740–3735[p]	14 × [p]

[a] Comparison of the rate of reassociation of macronuclear DNA with the rate of reassociation of *E. coli* DNA. All values were corrected for differences in GC content.
[b] Gillis et al. (1970).
[c] Gorovsky (1980).
[d] Gibson and Martin (1971).
[e] Cummings (1975).
[f] McTavish and Sommerville (1980).
[g] Cullis (1972, 1973).
[h] Steinbrück et al. (1981).
[i] Ammermann and Schlegel (1983).

[j] In these species the DNA content of both egg-shaped macronuclear structures together is considered as "macronuclear DNA content", since both parts behave as one functional unit in replication and division (see also p. 136).
[k] Not determined.
[l] Lauth et al. (1976).
[m] Dawson and Herrick (1982).
[n] Ammermann and Muenz (1982).
[o] Ammermann (unpubl.).
[p] Pelvat and DeHaller (1976).
[+] Mean, calculated from values which were determined by different authors.

explanation of the C-value paradox. According to this hypothesis, nuclear DNA has in addition to the well-known genic function a so-called nucleotypic function. Cavalier-Smith assumes that in large cells the transport of mRNA across the nuclear membrane may become a limiting factor for cellular growth. The rate of transport can only be increased by increasing the number of nuclear pores, which can be achieved by extending the surface of the nuclear membrane. If, as Cavalier-Smith believes, the area of the nuclear membrane is determined by the DNA content, then larger cells can only evolve by selection for increased DNA content. Thus, evolution might favor the accumulation in large cells of nongenic DNA (called skeletal DNA or S-DNA) which is assumed to consist mainly of repetitive sequences.

At first glance, the situation found in ciliates seems to be in excellent agreement with the skeletal DNA hypothesis. In fact, DNA elimination of hypotrichous ciliates has been cited by Cavalier-Smith (1978) as evidence supporting his hypothesis. He assumed that the elimination of repetitive sequences during nuclear differentiation removes S-DNA.

However, the facts are not in accordance with his assumption; at least a substantial modification of this hypothesis would be necessary. The reasons are based on the different functions of micro- and macronuclei. Micronuclear C-values are not correlated with cell size (see above). Micronuclei are transcriptionally nearly inactive (see Sects. 4.1.6 and 4.3.3) and have few nuclear pores. No mRNA transport across micronuclear membranes has been observed so far. On the other hand, macronuclei of hypotrichs are very DNA-rich but they contain only genic DNA. Their nuclear membranes show large numbers of nuclear pores arranged in specific patterns, but the membranes are not saturated with pores (Bardele, pers. commun.). Only macronuclear DNA content of hypotrichs is correlated with cell size, as mentioned above.

It should not entirely rejected that nuclear DNA in eukaryotes may have a "nucleotypic" function in addition to their genic function. In ciliates, however, only macronuclear DNA can fulfill such a function for the reasons indicated above. At least in hypotrichs no extra S-DNA consisting of repetitive sequences would be required since the amplified genic DNA might fulfill such an additional function. For all these reasons, the eliminated sequences of micronuclear genomes cannot be interpreted as being S-DNA.

In summary, the nuclear situation found in ciliates does not fit into the framework of the skeletal DNA hypothesis in its original version. Therefore, the hypothesis cannot provide a conclusive explanation of the C-value paradox of micronuclear genomes of ciliates.

8.1.3 The Selfish DNA Hypothesis

Another idea, the selfish DNA hypothesis, which aims at an explanation of the C-value paradox, has recently received much attention and provoked a controversial discussion (Doolittle and Sapienza 1980; Orgel and Crick 1980; Dover 1980; Cavalier-Smith 1980; Orgel et al. 1980; Dover and Doolittle 1980). The selfish DNA hypothesis contains the following idea: DNA of higher organisms falls into two classes, a specific one which fulfills the classic genetic functions, and a nonspecific class which fulfills a hitherto unrecognized function. The nonspecific, selfish, or parasitic DNA shall have two properties (Orgel and Crick 1980):

1) It arises when a DNA sequence spreads within a genome by forming additional copies of itself.
2) It makes no specific contribution to the phenotype.

The hypothesis has the serious drawback that it seems very problematic to falsify or support it by experimental evidence. Nevertheless, it shall be discussed whether it could give a plausible explanation of the biological function of nuclear differentiation phenomena or of the micronuclear C-value paradox in ciliates.

If existent, selfish DNA must occur within the micronuclear germ line genome. Since per definitionem the only "function" of selfish DNA is its multiplication and survival in genomes, this must occur in the germ line genomes. Macronuclear genomes are renewed after conjugation in most species (see Chap. 2). Therefore, a spreading of selfish DNA in macronuclear genomes would be without any significance for the survival of selfish sequences.

Repetitive sequences which might be selfish DNA are found within the micronuclei of all ciliates which were investigated so far. The fact that during nuclear differentiation repetitive sequences are eliminated nearly totally in hypotrichs (see Sect. 5.3.1) and to a lesser extent in other species (see Sect. 4.1.2) would not influence the interpretation of those sequences as parasitic or selfish DNA. Also the varying amounts of repetitive sequences in different species of hypotrichs as well as their sequence divergence between populations of one species are in accordance with this concept (cf. Sect. 5.3.1 and Steinbrück et al. 1981). But, at least in *Oxytricha* and *Stylonychia*, not only repetitive sequences, but also large amounts of unique sequences are eliminated during macronuclear development (Sect. 5.3.1). Greater amounts of unique sequences do not fit into the concept of selfish DNA. Even with respect to the additional assumption that repetitive selfish DNA sequences degenerated to unique sequences, an essential attribute of the selfishness would have been lost – the capacity of spreading throughout a genome by forming additional copies of itself.

It is noteworthy that at least for some of the micronucleus-restricted sequences other functions could be detected recently (Sects. 4.1.6, 4.3.3).

It is, therefore, not possible to conclude that the existence of differing amounts of selfish DNA sequences is the only basis of the micronuclear C-value paradox in ciliates. It is also not justified to assume that DNA elimination during nuclear differentiation removes only selfish DNA, even if some of the eliminated sequences might have the postulated properties of selfishness.

Also the closely related explanation that large parts of the micronuclear genomes are "junk", "useless", or "nonsense" DNA does not provide a reasonable and sufficient explanation of the micronuclear C-value paradox. In addition to the considerations mentioned above, it seems very unlikely that in some species more than 90% of the genomic sequences should make no contribution to the organismic phenotype. More important, there is no experimental possibility to confirm "nonfunctionality" of sequences.

8.1.4 Elimination of Pseudogenes and Introns

The elimination of intervening sequences or of pseudogenes during nuclear differentiation might be another function which is related to the C-value paradox. One may agree with Gall (1981) who states "that major differences in C-value will probably not be directly ascribable to differences in the number and size of introns". But if it could be shown that introns were eliminated during nuclear differentiation in ciliates that process could perhaps contribute to an explanation of the differences in C-values and it could be one additional function of nuclear differentiation.

Macronuclear development of hypotrichs shows the greatest extent of DNA elimination (Sect. 5.2). No true introns could be found in their genomes, but only a small number of genes have been analyzed so far (Sect. 5.3.3). True introns are intervening sequences which "split" the coding sequence of a gene and which are excised from the primary transcript by a precise RNA splicing process. An entirely new type of small introns was detected recently in micronuclear genes of *Oxytricha* by Klobutcher et al. (1984, see Sect. 5.3.3). These internal eliminated sequences (IES) differ from "true introns" (see above), since they are eliminated from the transcriptionally inactive micronuclear genome by a *DNA* splicing process. This extraordinary case demonstrates that small introns of a very special type are eliminated during macronuclear development in at least one hypotrichous species. However, the vast majority of eliminated sequences in hypotrichs seems to consist of intergenic spacer sequences (Prescott 1983).

One other case of the existence of introns in ciliates has been reported. Wild and Gall (1979) found an intervening sequence of about 400 bp in the extrachromosomal macronuclear rDNA of one strain of *Tetrahymena pigmentosa*. Another strain of the same species lacks that intron. The absence or presence of that intron has obviously no functional importance.

One may conclude from these two cases only that elimination of a special type of intron does occur in some species, but that, on the other hand, true introns may also occur in macronuclear genomes of ciliates. Therefore, elimination of introns might be one function of nuclear differentiation in some species. But it does not seem to be the general case concerning all species and all introns. It cannot explain the large differences in micronuclear C-values.

The occurrence of pseudogenes has been reported for many groups of eukaryotic organisms. Although their function and mode of generation in cells is not clear, the following assumption seems to be justified. Pseudogenes may play an important role in the generation of multigene families which among other functions increase genome flexibility for adaptational requirements (Proudfoot 1980). If pseudogenes indeed contribute to the evolutionary flexibility of a species, they should be found in micronuclear genomes of ciliates. But they can be assumed to be dispensable for macronuclear genomes, since they are not transcribed. Pseudogenes might, therefore, contribute to the micronuclear C-value paradox and they might belong to those sequences which are eliminated in those species which do eliminate or are underreplicated perhaps in others.

It is especially interesting under this aspect that Lee et al. (1983) predict from their study of processed β-tubulin pseudogenes in man that pseudogenes of the processed type should be more easily generated in cells with a high level of transcription. In higher organisms such a way of generating pseudogenes by some sort of reverse transcription and integration of cDNA into the genome would be stable only if the process occurs in germ line cells.

In ciliates the germ line nuclei – the micronuclei – lie within a cytoplasm which contains a high level of RNA transcripts (Nock 1981). Thus, according to this hypothesis ciliate micronuclei would, therefore, be excellently suited for accumulating large amounts pseudogenes of the processed type.

Do the experimental data found in ciliate nuclear differentiation, however, support such speculations? Multigene families have been found in several ciliate species. Histone genes, actin genes, α- and β-tubulin genes, as well as a number of genes of unknown functions form gene families in macronuclear genomes of several ciliate species (see Sects. 4.1.5.2, 5.3.3). The situation of micronuclear genomes is known only in a few cases. Whether the known gene families contain pseudogenes is unknown. Most of them contain only two to three members. The important question remains unanswered whether micronuclear genomes contain larger gene families. Nor is there indirect evidence for the existence of larger numbers of pseudogenes in micronuclear genomes. The juxtaposition of poly-(A) tracts close to single members of a micronuclear gene family would provide such indirect evidence.

So far, no conclusive evidence can be given that accumulation of pseudogenes contributes to the C-value paradox or that elimination or underrepresentation of pseudogenes is a facet of nuclear differentiation in ciliates. However, the hypothesis might be confirmed by sequence analyses in the near future.

8.2 Nuclear Differentiation and Gene Activation

The most impressive differences between micro- and macronuclei are the different appearances and the different activities of the two kinds of nuclei. The micronuclei are small, condensed, and largely inactive nuclei, the macronuclei are very active in transcription and contain largely unfolded chromatin with only a few heterochromatin areas in most species (see Chap. 2). Therefore, it seems possible that nuclear differentiation is involved in gene activation in ciliates. The main question is: is there any experimental evidence for mechanisms which could activate the transcriptionally largely inactive micronuclear germ line genome?

Since it has been shown that DNA sequence rearrangements may be connected with gene activation in some eukaryotes (e.g., Nasmyth 1982; Borst and Cross 1982; Tonegawa 1983), it is tempting to suggest that comparable molecular mechanisms operating during nuclear differentiation in ciliates may have similar functions. At present, the experimental data do not allow discussion of such a regulatory function for ciliates in general. But nevertheless, recent data seem to point in that direction.

Yao et al. (1984) found evidence for rearrangements during macronuclear development in *Tetrahymena* which would allow such an interpretation (see Sect. 4.1.5). Germ line specific sequences are eliminated by a breakage and rejoining mechanism during nuclear differentiation. This DNA splicing process removes sequences which are adjacent to the sequences which are maintained in the somatic macronucleus. Whether these rearrangements concern, in fact, control sequences necessary for the suppression of gene activities must be tested in further experiments using the DNA fragments involved. The rearrangements are estimated to occur with a high frequency in the *Tetrahymena* genome (Sect. 4.1.5) and might, therefore, play a crucial role in activating the inert germ line genome.

Similarly, Callahan et al. (1984) found indications for rearrangements during macronuclear differentiation involving the sequences flanking both sides of the α-tubulin gene of *Tetrahymena*. But, on the other hand, histone H4 gene expression in *Tetrahymena* does not seem to be linked with rearrangements during macronuclear development (Bannon et al. 1984). At least large rearrangements comparable to those found in yeast mating type genes, immunoglobulin genes of mammals, or surface glycoprotein genes in *Trypanosomes* (l.c.) must be ruled out for the histone H4 genes of *Tetrahymena*, but very small or distant rearrangements cannot be ruled out.

Just such very small changes have been found in the hypotrich *Oxytricha* (Klobutcher et al. 1984, see Sects. 5.3.3, 8.1.4). The elimination of small IES elements from the germ line genome of *Oxytricha* seems to be comparable to the elimination of larger fragments of the micronuclear genome of *Tetrahymena* (see Sect. 4.1.5) insofar as in both cases presumably a DNA splicing process is involved. But in the hypotrich *Oxytricha* the eliminated sequences lie within a gene sequence, in *Tetrahymena* only gene flanking sequences seem to participate in the elimination-ligation process. It was mentioned in the same study that genomic hybridization experiments revealed that similar splicing processes can also be assumed to occur in other genes of *Oxytricha*.

The presently available experimental data do not allow conclusive evidence whether sequence rearrangements or splicing processes are essential requirements of gene activation during macronuclear development in ciliates.

8.3 Differential Regulation of Replication

There are other regulatory implications of nuclear differentiation in ciliates which must be considered. S-phases of macro- and micronuclear genomes never coincide exactly in ciliates. S-phases of micronuclei are always shorter than those of the corresponding macronuclei (see compilation in Raikov 1982). This suggests that the regulation of replication is changed during macronuclear development. In any case, when nuclear development is connected with extrachromosomal amplification (e.g., rDNA in *Tetrahymena*, Sect. 4.1.5.1 or *Paramecium* species, Sect. 4.3.2) or with more or less extensive fragmentation and elimination (e.g., *Tetrahymena*, Sect. 4.1.5, *Glaucoma*, Sect. 4.2, hypotrichs, Sect. 5.3) additional problems with regards to the regulation of replication arise. Every independent piece of DNA must have at least one origin of replication if it should not be lost during subsequent nuclear divisions or underrepresented when amplification takes place. It could be demonstrated that extrachromosomal rDNA molecules of *Tetrahymena* start with replication in the center of the large palindromes (Truett and Gall 1977; Kiss et al. 1981; see also Sect. 4.1.5.1 and Fig. 4). The gene-sized DNA pieces start replication near one or both ends of every macronuclear DNA molecule in *Oxytricha* (Murti and Prescott 1983) and *Stylonychia* (Steinbrück, unpubl.).

Another important regulatory problem seems to be connected with the amplification processes during nuclear differentiation (see Chap. 2, Sects. 4.1.5, 5.2, 8.5). Every independent DNA molecule of late macronuclear anlagen in hypotrichs and, presumably, amplified sequences of other ciliates also must, however, "know" how many rounds of replication must take place during nuclear development to give the final characteristic macronuclear gene copy number (see also Fig. 7). The presumed mechanism would not be necessary for the integrated micronuclear copies and should be inactivated there. So far no information is available on starting points of replication in micronuclei of hypotrichous or other ciliates. It has only been demonstrated that mutations can occur in micronuclei of hypotrichs which lead to overamplification of the mutated gene in the macronuclei. As a result, a suspicious deviation of the species-specific macronuclear

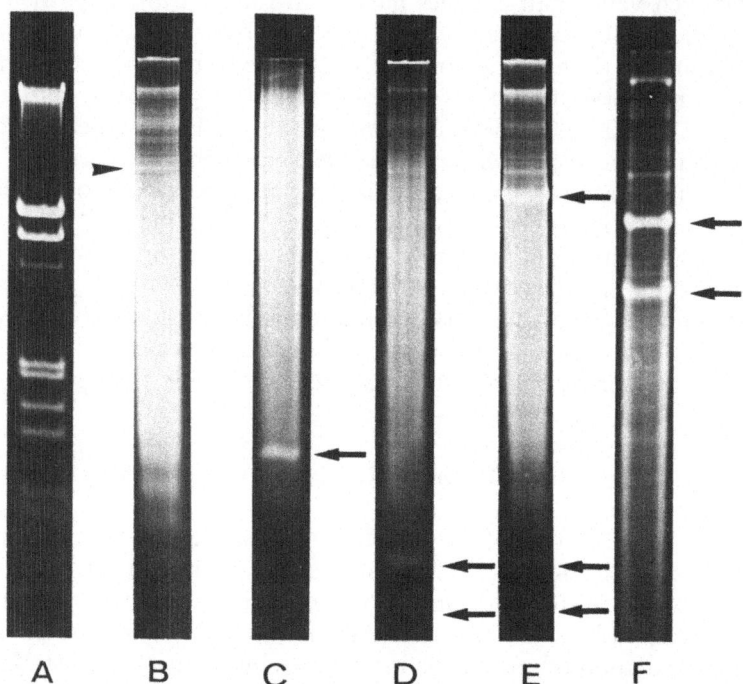

Fig. 8. Overamplification of individual DNA sequences in macronuclear genomes of hypotrichous ciliates. Macronuclear DNA of three strains of *Stylonychia lemnae* and of one strain of *Euplotes vannus* was separated according to size by electrophoresis on a 1% agarose gel. After electrophoresis the gel was stained with ethidium bromide and photographed under UV illumination. *Lane A* shows the molecular weight marker, λ-DNA, digested with Eco RI and Hind III. *Lanes B–E* show macronuclear DNA of different strains of *S. lemnae*, one exhibiting no overamplification (*lane B*), and others with one, two, or three overamplified sequences (*arrows*). *Lanes D* and *E* show macronuclear DNA isolated from the same strain at different clonal ages. The *arrowhead* marks the rRNA genes which occur in higher copy numbers in all hypotrichous species investigated so far . *Lane F* shows macronuclear DNA of a strain of *E. vannus* containing two overamplified sequences. From Steinbrück (1983)

gene copy number can be observed (Steinbrück 1983). Examples of overamplified genes are shown in Fig. 8. The characteristic of overamplificating a certain gene can be inherited via the germ line genome to the next generation after conjugation. Sequence comparisons of mutated and wild-type genes are in progress. They could perhaps help to unravel the problem of regulating gene copy numbers in macronuclear genomes, at least for hypotrichs.

Closely related to fragmentation of genomes during macronuclear development and to extrachromosomal amplification is another regulatory problem which has been discussed earlier (Sects. 4.1.5.1, 4.1.5.3, 4.2, 5.3.2). Every newly generated piece of DNA, except circles, must contain telomere structures which allow the replication of the ends of the molecules and thereby their maintenance. In *Tetrahymena, Glaucoma, Paramecium,* as well as in hypotrichs, telomere structures have been analyzed. In all cases it must be assumed that the telomere structures are added to the DNA fragments by rearrangement or by step-by-step addition processes (reviewed by Blackburn et al. 1983).

8.4 Other Putative Regulatory Functions

Quite another regulatory mechanism connected with nuclear development was recently found in *Paramecium tetraurelia* (Epstein and Forney 1984) (see Sect. 4.3.3). A mutation was detected which seems to affect the expression of the gene coding for the immobilization antigen A. The authors suggest that their data provide the first evidence that mechanisms exist in ciliates to control the expression of a gene by regulating its incorporation into the developing macronuclei. The conclusion that the detection of that mutant also indicates the presence of a regulatory mechanism operating in the same way in nonmutated cells, seems to be premature. But if this can be supported by further experimental evidence from other genes, in fact, a novel mechanism for regulating gene expression by means of differential transmission of genes from a germ line genome to a somatic genome would have been found in ciliates.

Further differences between macro- and micronuclear genomes might be implicated in regulatory functions of nuclear differentiation. It was mentioned before (Sect. 5.3.2) that macronuclei of *Stylonychia* react in contrast to micronuclei and anlagen with antibodies specific for left-handed Z-DNA. It was discussed by the authors (Lipps et al. 1983) that macronuclear DNA might contain a greater negative superhelical density which stabilizes Z-conformation. Since it is known that greater negative superhelical density favors transcription, the detection of Z-DNA conformation in ciliate macronuclei might give an indication of a mechanism of transcriptional activation.

Also the differences found in nuclear proteins and chromatin structure of both kinds of nuclei (see Sects. 4.1.3, 5.3.4) are highly likely to be correlated with the differences in transcriptional activity. But the functional analysis of chromatin structure and chromatin proteins has just been started in ciliates. A further discussion of putative regulatory functions of these characteristics must, therefore, await new data obtained by functional tests and not only by comparisons of the underlying structures.

8.5 Putative Functions of Giant Chromosomes

Closely related with possible regulatory functions of the molecular mechanisms acting during nuclear differentiation might be the formation of giant chromosomes found in macronuclear anlagen of some species of ciliates (see Sects. 5.2,
6.2, 6.4). Giant chromosomes have been detected in macronuclear anlagen of hypotrichous species belonging to the genera *Stylonychia* (see Fig. 6) (Ammermann
1965; Alonso and Perez-Silva 1965), *Euplotes* (Rao and Ammermann 1970), and
Oxytricha (Spear and Lauth 1976). Besides the order Hypotrichida, giant chromosomes have been found in *Nyctotherus*, Heterotrichida (Sect. 6.4) (Golikova
1965), and *Chilodonella*, Cyrtophorida (Sect. 6.2) (Radzikowski 1973). Despite
some structural similarities, e.g., alternating bands and interbands, dipteran giant
chromosomes and those of ciliates show striking differences in ultrastructure
(Meyer and Lipps 1984) and seem to have entirely different functions. In contrast
to the considerable transcriptional activity of dipteran giant chromosomes, no
really convincing evidence in ciliate giant chromosomes has been presented so far.
Despite the fact that this point has been discussed controversially (see Sect. 5.3,
Alonso and Jareno 1974; Ammermann et al. 1974), it seems to be now clear that
differential transcription is not a main function of giant chromosomes in macronuclear anlagen of ciliates. The question remains open which other functions can
be taken into account. Several observations reveal that differential replication occurs in different segments of giant chromosomes of *Stylonychia*, *Oxytricha*, and
Chilodonella (Ammermann et al. 1974; Spear and Lauth 1976; Radzikowski
1979). There is also detectable a reduction of satellite DNA content during the
formation of giant chromosomes in *Stylonychia* and *Oxytricha* and a shift in
buoyant density of the polytene stage DNA in *Oxytricha* (Ammermann et al.
1974; Spear and Lauth 1976). One may conclude from these data that the formation of giant chromosomes during nuclear differentiation may be a prerequisite
for differential amplification of genes. The over- or underreplication of individual
bands or interbands may result in the macronuclear gene copy numbers which differ from gene to gene and are specific for any gene (see Fig. 7, for example) in hypotrichs. The species-specific macronuclear DNA banding patterns (Swanton et
al. 1980 b; Steinbrück et al. 1981) would be the result of that process. Such a mechanism would provide without any additional transcriptional regulatory system a
wide range of different transcript concentrations within a cell.

Another reason for the existence of giant chromosomes may be found in the
elimination process (cf. Sect. 5.3 and Fig. 5). Polytenization and formation of
bands and interbands might be necessary for an exact cutting and/or rearrangement process which finally results in a functioning macronuclear genome. But one
should keep in mind that ciliate species exist which exhibit fragmentation (e.g.,
Glaucoma, Sect. 4.2) or elimination and rearrangement of sequences (e.g., *Tetrahymena*, Sect. 4.1.5) and which obviously do not form giant chromosomes during
macronuclear development. Only a more detailed analysis of the elimination processes can provide evidence as to whether basically the same or rather different
molecular mechanisms have been evolved in those groups of ciliates.

8.6 Protection Against Genetic Damages

The high macronuclear DNA amounts raise some problems in that the generation and replication of large somatic genomes requires a huge supply of cellular energy and materials. The drastic elimination of nearly all those sequences of the germ line genome which are not essential for somatic functions during nuclear differentiation in hypotrichs (Sect. 5.2) may therefore have been for reasons of cellular economy.

More importantly, numerous rounds of replication during macronuclear development may lead to an accumulation of replication errors. This may be true even in cells with an active repair system (e.g., Dittmann 1978). So far there is no definitive evidence for an accumulation of mutations in macronuclear genomes. The sequence heterogeneity, however, of some macronuclear genes in *Oxytricha* (Klobutcher et al. 1984) or of α-tubulin genes of *Stylonychia* (Helftenbein 1985) may be interpreted in that way.

In addition, the very active macronuclear genomes containing mainly extended chromatin may be a better target for damages produced by environmental influences, e.g., irradiation, than the small, inactive micronuclear genomes which contain varying amounts of sequences which seem to be dispensable for vegetative functions. However, there is some contradictory evidence regarding this point. It is well-known that ciliate cells show a great resistance against irradiation damage. For example, cells of *Stylonychia* tolerate X-ray doses of more than 35 kr without detectable disadvantages (Dittmann 1978) and some cells survive even 1,000 kr (Ammermann, pers. commun.). One may conclude from this contradictory evidence that the disadvantages of the large somatic genomes are balanced by advantages resulting from the specific macronuclear genome organization. The tolerance for damages may be based on the amplification of essential genes in some species, or of the whole genomes in others. High gene copy numbers may provide a way for macronuclear genomes to tolerate a certain amount of damage or of inactivated gene copies. Of course, such a tolerance is only possible if the actual copy number of a gene is in excess of that number which is absolutely necessary for normal cell function. A molecular mechanism capable of regulating macronuclear gene copy numbers (see Sect. 8.3) would be required. The tolerance would be increased by low molecular weight (cf. Hutchinson 1961) of macronuclear DNA as it is found in hypotrichs (Sect. 5.3.1) and in some others.

Another macronuclear character seems to be closely related to that problem. The lack of a mitotic or meiotic apparatus in macronuclei as well as the underrepresentation of repetitive sequences in some species (e.g., *Tetrahymena*, Sect. 4.1.3) or the elimination of most of them in others (e.g., hypotrichs, Sect. 5.3) might prevent mechanisms which are otherwise essential for intragenomic exchange or correction processes. Only conjugation or autogamy can lead to rejuvenescence. More or less defective macronuclear genomes are replaced by new ones starting from the better protected germ line genomes. Micronuclei, however, have a mitotic apparatus and contain repetitive sequences. They seem to be better protected against damages due to their tightly packaged, inactive genetic material. However, double strand breaks would quickly lead to genetically defective or even lethal cells due to the action of the mitotic apparatus. Single base exchanges, however, and other minor mutations might be accumulated in their genomes often without any harm since large amounts of their DNA seems to be dispensable for most vegetative functions. This might be especially true for micronuclei with large C-values (cf. Sect. 8.1).

In summary, counteracting evolutionary forces seem to have led to the separation of functions between macro- and micronuclei. Enlarged evolutionary flexibility as well as functional versatility of ciliate genomes seem to be the important consequences of their nuclear dualism.

9 Are Differences Between Germ Line and Soma Genomes in Ciliates of General Importance for Comparable Phenomena in Multicellular Eukaryotes?

The evolutionary relationships between ciliates and multicellular eukaryotes are far from being clear (see Discussion in Sect. 7.1). Undoubtedly both ciliates and multicellular eukaryotes have taken separate evolutionary pathways for very long times. Therefore, some suspicious similarities in the organization of macro- and micronuclear genomes of ciliates with peculiarities in the organization of germ line and somatic genomes of other eukaryotes may be based on convergency. This means the similarities were evolved independently due to similar functional constraints. Nevertheless, some of the underlying molecular mechanisms might reflect a common root of nuclear differentiation in most eukaryotes. Their elucidation in one group would be relevant for the understanding of similar mechanisms in other groups. But also if molecular mechanisms which are involved in nuclear differentiation were evolved independently, the comprehension of their functions can often contribute essential aspects to the understanding of the convergent processes.

Some of the molecular mechanisms which can be studied especially easy and comfortably in ciliates and whose elucidation might contribute to the understanding of similar events in other eukaryotes shall be summed up in this chapter.

One advantage of the study of nuclear differentiation in ciliates which must be mentioned first seems very obvious. Whereas germ line and soma genomes of multicellular organisms are always situated in different cells, they lie within the same cytoplasm in ciliates. This unique situation found in ciliates (and foraminiferans) rules out many potential influences on nuclear differentiation which might be based on differences of the surrounding cytoplasmic environment.

Another advantage is the large size of some ciliates. Large ciliate cells can be handled conveniently, for instance, in studies using microinjection or other micromanipulation techniques. Furthermore, many ciliate species can be produced in large quantities starting with a single cell. Therefore, working with ciliate cells provides many of the advantages of microbiological techniques which favor otherwise the study of prokaryotes.

The generation of extrachromosomal DNA molecules during nuclear differentiation in ciliates seems to be based on mechanisms whose study is relevant for the understanding of similar events in other eukaryotes. The generation of macronuclear rDNA molecules in *Tetrahymena* species (see Sect. 4.1.5.1) belongs to the best-studied cases of the generation of extrachromosomal DNA. The development of gene-sized macronuclear DNA molecules in hypotrichs (see Sect. 5.3.1) and of small-sized macronuclear DNA in other ciliates (Sects. 4.2, 4.3.2, 6.3) seem to be based, at least partially, on similar mechanisms as in the well-studied rDNA generation in *Tetrahymena*. The study of DNA fragmentation, elimination, and amplification has been intensified considerably during the last few years. The results obtained are surely relevant for similar, but so far less intensively, studied cases in other eukaryotes, e.g., *Cyclops* or *Ascaris* (see Chaps. Tobler and Hennig, this volume).

Especially important is the study of telomer structure and function in ciliate DNA molecules (Sects. 4.1.5, 4.2, 5.3.2). Linear DNA molecules require special telomer structures at their ends in order to be replicated correctly. It must be assumed that the problem of telomer replication is common to all chromosome ends found in eukaryotic genomes. But in most genomes with high complexities and high molecular weight DNA, the study of telomer structures, e.g., their formation and function, is severely hampered due to the low concentration of ends in those genomes. Suitable vectors for cloning linear DNA molecules have been developed only recently for eukaryotic systems and they have some limitations (e.g., Szostak and Blackburn 1982; Dani and Zakian 1983). The gene-sized linear DNA molecules found in macronuclear genomes of hypotrichs as well as the easily obtainable rDNA molecules of *Tetrahymena* and perhaps of other species, too (cf. Fig. 4) offer especially suitable material for such important investigations (Pluta et al. 1984). The fact that very similar telomere structures have been detected not only in different ciliate species, but also in genomes of slime moulds, yeast, and trypanosomes (reviewed by Blackburn et al. 1983; see also Blackburn and Challoner 1984) supports the view that the study of telomeres in ciliates may be relevant for the understanding of formation and function of end structures of linear DNA molecules in general.

The elimination of DNA sequences during nuclear differentiation is found among ciliates more often than in any other group of organisms. Therefore, ciliates provide the best opportunity for comparative studies of DNA elimination processes and of detailed analyses of the eliminated sequences. Perhaps such studies might contribute to better understanding of DNA elimination processes and to a deeper insight into the significance of over- and underreplication processes found frequently in many other groups of eukaryotic organisms.

It must be added that not only sequence elimination, but also the elimination of whole chromosomes, has been observed during nuclear differentiation in ciliates (see Sect. 5.2). The elimination of pycnotic chromosomes during an early stage of macronuclear development closely resembles the elimination of B-chromosomes in different groups of organisms (see Chap. Gerbi, this volume). Since macro- and micronuclear genomic libraries are now available on *Stylonychia*, the study of those sequences which are contained in the eliminated chromosomes, which might be B-chromosomes, is now possible. Perhaps such a study of B-chromosomal DNA sequences would promote an understanding of B-chromosomes which have not been thoroughly investigated at the molecular level in other organisms.

Comparisons of macro- and micronuclear proteins (cf. Sects. 4.1.3 and 5.3.4) in ciliates provide another favorable method of examining the problem of nuclear differentiation. Both sets of proteins which contribute to the differences found between the two kinds of nuclei are products of the activity of the somatic genome. The study of the mechanisms of separating and directing the components of both genomes into their destinated compartments seems to be a very promising step toward the understanding of the differences between germ line and soma genomes. No other organisms, except ciliates with their nuclear dualism, seem so well-suited for such studies which are relevant to all eukaryotes.

The gene-sized DNA molecules of hypotrichous macronuclear genomes as well as macronuclear DNA fragments of other ciliates (e.g., *Glaucoma*, Sect. 4.2) can be used for studies of another important problem. Most of the macronuclear DNA molecules of hypotrichs which have been analyzed so far in detail contain, in addition to the protein coding region, only short flanking sequences (see Fig. 4 and cf. Sect. 5.3.3).

If trans-acting regulatory sequences are necessary and responsible for a coordinated expression of different genes which lie on independent DNA molecules, e.g., the α- and β-tubulin coding sequences in hypotrichs, their detection and functional analysis should be greatly facilitated in macronuclear genomes of hypotrichous ciliates. The problem of coordinated expression of genes which are not clustered, but located on widely separated positions of one chromosome or on different chromosomes in eukaryotes, is so far poorly understood in most cases.

In addition, comparisons of these noncoding sequences of macronuclear genes of ciliates with the corresponding flanking regions in the micronuclear genomes might provide some insight into the problem of inactivation of the germ line genomes. Especially the detection of the internal eliminated sequences (IES, see Sect. 5.3.3) by Klobutcher et al. (1984) is very promising and might bring a breakthrough in this field.

Finally, micronuclear genomes of ciliates show a wide range of C-values as discussed above (Chap. 8). Comparative studies of genomes which differ in their C-values in connection with known data on the reduction of C-values during nuclear differentiation in some groups of ciliates provide an entirely new direction of investigation of the hitherto unresolved C-value paradox. Thus, success in understanding the problem in ciliates would also be relevant for other organisms.

Acknowledgments. I am grateful to D. Ammermann, C.-F. Bardele, M. Schlegel, H.-M. Seyffert, and the editor, W. Hennig, for critical comments and suggestions to the manuscript. I thank M. Schlegel and D. Ammermann for kindly providing photographs and unpublished data.

References

Allis CD, Glover CVC, Gorovsky MA (1979) Micronuclei of Tetrahymena contain two types of histone H3. Proc Natl Acad Sci USA 76:4857–4861

Allis CD, Bowen JK, Abraham GN, Glover CVC, Gorovsky MA (1980a) Proteolytic processing of histone H3 in chromatin: a physiologically regulated event in Tetrahymena micronuclei. Cell 20:55–64

Allis CD, Glover CVC, Bowen JK, Gorovsky MA (1980b) Histone variants specific to the transcriptionally active, amitotically dividing macronucleus of the unicellular eucaryote, Tetrahymena thermophila. Cell 20:609–617

Allis CD, Ziegler YS, Gorovsky MA, Olmsted JB (1982) A conserved histone variant enriched in nucleoli of mammalian cells. Cell 31:131–136

Allis CD, Wiggins JC, Richman R, Chicoine L, Wenkert D (1984) Histone rearrangements accompany nuclear differentiation and dedifferentiation in Tetrahymena thermophila. In: Abstracts of papers presented at the meeting on "Ciliate molecular genetics" 2.5.–6.5.1984. Cold Spring Harbor Laboratory, New York

Alonso P, Jareno MA (1974) Incorporacion de uridina-H³ en el esbozo macronuclear de Stylonychia mytilus. Microbiol Esp 27:199–211

Alonso P, Pérez-Silva J (1965) Giant chromosomes in protozoa. Nature (Lond) 205:313–314

Amabis JM (1983) DNA-puffing patterns in the salivary glands of Trichosia pubescens (Diptera, Sciaridae). Genetica (The Hague) 62:3–13

Ammermann D (1965) Cytologische und genetische Untersuchungen an dem Ciliaten Stylonychia mytilus Ehrenberg. Arch Protistenkd 108:109–152

Ammermann D (1970) The micronucleus of the ciliate Stylonychia mytilus, its nucleic acid synthesis and its function. Exp Cell Res 61:6–12

Ammermann D (1971) Morphology and development of the macronuclei of the ciliates Stylonychia mytilus and Euplotes aediculatus. Chromosoma (Berl) 33:209–238

Ammermann D (1982) Mating types in Stylonychia mytilus Ehrenberg. Arch Protistenkd 126:373–381

Ammermann D, Muenz A (1982) DNA and protein content of different hypotrich ciliates. Eur J Cell Biol 27:22–24

Ammermann D, Schlegel M (1983) Characterization of two sibling species of the genus Stylonychia (Ciliata, Hypotricha): S. mytilus Ehrenberg, 1838 and S. lemnae n. sp. I. Morphology and reproductive behavior. J Protozool 30:290–294

Ammermann D, Steinbrück G, v Berger L, Hennig W (1974) The development of the macronucleus in the ciliated protozoan, Stylonychia mytilus. Chromosoma (Berl) 45:401–429

Ammermann D, Steinbrück G, Baur R, Wohlert H (1981) Methylated bases in the DNA of the ciliate, Stylonychia mytilus. Eur J Cell Biol 24:154–156

Ax P (1966) Die Bedeutung der interstitiellen Sandfauna für allgemeine Probleme der Systematik, Ökologie und Biologie. Veröff Inst Meeresforsch Bremerhaven, Suppl 2:15–66

Balbiani EG (1890) Étude sur le Loxode. Ann Micrograph 2:401–431

Bannon GA, Bowen JK, Yao M-C, Gorovsky MA (1984) Tetrahymena H4 genes: structure, evolution and organization in macro- and micronuclei. Nucleic Acids Res 12:1961–1975

Berger JD (1973) Nuclear differentiation and nucleic acid synthesis in well-fed exconjugants of Paramecium aurelia. Chromosoma (Berl) 42:247–268

Blackburn EH (1982) Characterization and species differences of rDNA: protozoa. In: Busch H, Rothblum L (eds) The cell nucleus, vol 10. Academic Press, New York, p 145–170

Blackburn EH (1984) Telomers: Do the ends justify the means? Cell 37:7–8

Blackburn EH, Challoner PB (1984) Identification of a telomeric DNA sequence in Trypanosoma brucei. Cell 36:447–457

Blackburn EH, Gall JG (1978) A tandemly repeated sequence at the termini of the extrachromosomal ribosomal RNA genes in Tetrahymena. J Mol Biol 120:33–53

Blackburn EH, Szostak JW (1984) The molecular structure of centromeres and telomeres. Annu Rev Biochem 53:163–194

Blackburn EH, Budarf ML, Challoner PB, Cherry JM, Howard EA, Katzen AL, Pan W-C, Ryan T (1983) DNA termini in ciliate macronuclei. Cold Spring Harborg Symp Quant Biol 47:1195–1207

Bobyleva NN, Kudrjavtsev BN, Raikov IB (1980) Changes of the DNA content of differentiating and adult macronuclei of the ciliate, Loxodes magnus (Karyorelictida). J Cell Sci 44:375–394

Borchsenius SN, Sergejeva GI (1979) Characterization of DNA from the ciliate, Bursaria truncatella O. F. Müller, vegetative cells. Tsitologiya 21:327–333 (in Russian with English summary)

Borden D, Whitt GS, Nanney DL (1973) Electrophoretic characterization of classical Tetrahymena pyriformis strains. J Protozool 20:693–700

Borden D, Miller ET, Whitt GS, Nanney DL (1977) Electrophoretic analysis of evolutionary relationships in Tetrahymena. Evolution 31:91–102

Borst P, Cross GAM (1982) Molecular basis for trypanosome antigenic variation. Cell 29:291–303

Bostock CJ, Prescott DM (1972) Evidence of gene diminution during the formation of the macronucleus in the protozoan Stylonychia. Proc Natl Acad Sci USA 69:139–142

Boswell RE, Klobutcher LA, Prescott DM (1982) Inverted terminal repeats are added to genes during macronuclear development in Oxytricha nova. Proc Natl Acad Sci USA 79:3255–3259

Boswell RE, Jahn CL, Greslin AF, Prescott DM (1983) Organization of gene and non-gene sequences in micronuclear DNA of Oxytricha nova. Nucleic Acids Res 11:3651–3663

Brachet J, Bonotto S (eds) (1970) Biology of Acetabularia. Elsevier/North-Holland, Amsterdam

Brunk CF, Tsao SGS, Diamond CH, Ohshi PS, Tsao NNG, Pearlman RE (1982) Reorganization of unique and repetitive sequences during nuclear development in Tetrahymena thermophila. Can J Biochem 60:847–853

Bruns PJ (1984) Tetrahymena thermophila. In: O'Brien SJ (ed) Genetic maps 1984. Cold Spring Harbor Laboratory, New York, pp 211–215

Bruns PJ, Brussard TEB (1981) Nullisomic Tetrahymena: eliminating germinal chromosomes. Science (Wash DC) 213:549–551

Bütschli O (1876) Studien über die ersten Entwicklungsvorgänge der Eizelle, die Zellteilung und die Conjugation der Infusorien. Abhandl Senckenberg Naturforsch Ges 10:213–464

Callahan RC, Shalke G, Gorovsky MA (1984) Developmental rearrangements associated with a single type of expressed α-tubulin gene in Tetrahymena. Cell 36:441–445

Calos MP, Miller JH (1980) Transposable elements. Cell 20:579–595

Caplan EB (1975) A very rapidly migrating f1 histone associated with gene-sized pieces of DNA in the macronucleus of Oxytricha sp. Biochim Biophys Acta 407:109–113

Caplan EB (1977) Histones and other basic nuclear proteins in genetically active and genetically inactive nuclei of the ciliate, Oxytricha sp. Biochim Biophys Acta 479:214–219

Cartinhour SW, Herrick GA (1984a) Three different macronuclear DNAs in Oxytricha fallax share a common sequence block. Mol Cell Biol 4:931–938

Cartinhour S, Herrick G (1984b) Alternate juxtaposition of macronuclear sequences in Oxytricha fallax. In: Abstracts of papers presented at the meeting on "Ciliate molecular genetics" 2.5.–6.5.1984. Cold Spring Harbor Laboratory, New York, p 6

Cavalier-Smith T (1978) Nuclear volume control by nucleoskeletal DNA, selection for cell volume and cell growth rate, and the solution of the C-value paradox. J Cell Sci 34:247–278

Cavalier-Smith T (1980) How selfish is DNA? Nature (Lond) 285:617–618

Cavalier-Smith T (1982) Skeletal DNA and the evolution of genome size. Annu Rev Biophys Bioeng 11:273–302

Chen TT (1940) Polyploidy in Paramecium bursaria. Proc Natl Acad Sci USA 26:239–340

Cleffmann G (1968) Regulierung der DNS-Menge im Makronucleus von Tetrahymena. Exp Cell Res 50:193–207

Conner RL, Koroly MJ (1973) Chemistry and metabolism of nucleic acids in Tetrahymena. In: Elliott AM (ed) Biology of Tetrahymena. Dowden, Hutchinson and Ross, Stroudsburg, Pa

Corliss JO (1979) The ciliated protozoa. Pergamon, Oxford

Corliss JO, Hartwig E (1977) The „primitive" interstitial ciliates: Their ecology, nuclear uniqueness, and postulated place in the evolution and systematics of the phylum Ciliophora. Mikrofauna Meeresboden 61:65–88

Cullis CA (1972) The basis of cell-to-cell transformation in Paramecium bursaria. II. Investigation into the molecular nature of the transforming agent. J Cell Sci 11:611–619

Cullis CA (1973) DNA amounts in the nuclei of Paramecium bursaria. Chromosoma (Berl) 40:127–133

Cummings DJ (1975) Studies on macronuclear DNA from Paramecium aurelia. Chromosoma (Berl) 53:191–208

Cummings DJ, Tait A (1975) The isolation of nuclei from Paramecium aurelia. In: Prescott DM (ed) Methods in cell biology, vol 9. Academic Press, New York, pp 281–309

Czihak G (1964) Experiments on nuclear differentiation in the foraminifer Rotaliella heterokaryotica Grell by means of UV-irradiation. Exp Cell Res 35:372–380

Dani GM, Zakian VA (1983) Mitotic and meiotic stability of linear plasmids in yeast. Proc Natl Acad Sci USA 80:3406–3410

Dawson D, Herrick G (1982) Micronuclear DNA sequences of Oxytricha fallax homologous to the macronuclear inverted terminal repeat. Nucleic Acids Res 10:2911–2924

Dawson D, Herrick G (1984) Telomeric properties of C_4A_4-homologous sequences in micronuclear DNA of Oxytricha fallax. Cell 36:171–177

Diamond CH, Ohashi PS, Rose A, Tsao NNG, Pearlman E (1984) Analysis of a micronucleus limited sequence from Tetrahymena thermophila. In: Abstracts of papers presented at the meeting on „Ciliate molecular genetics" 2.5.–6.5.1984. Cold Spring Harbor Laboratory, New York, p 10

Dittmann F-N (1978) Untersuchungen zur Wirkung von UV- und Röntgenstrahlen und zur Reparatur der Strahlenschäden bei dem Ciliaten Stylonychia mytilus. Thesis Univ Tübingen

Doerder FP (1979) Regulation of macronuclear DNA content in Tetrahymena thermophila. J Protozool 26:28–35

Doerder FP, DeBault LE (1975) Cytofluorometric analysis of nuclear DNA during meiosis, fertilization and macronuclear development in the ciliate, Tetrahymena pyriformis. J Cell Sci 17:47–93

Doolittle WF, Sapienza C (1980) Selfish genes, the phenotype paradigm and genome evolution. Nature (Lond) 284:601–603

Dover G (1980) Ignorant DNA. Nature (Lond) 285:618–620

Dover G, Doolittle WF (1980) Modes of genome evolution. Nature (Lond) 288:646–647

Dragesco J (1960) Ciliés mésopsammique littoraux. Systematique morphologie, écologie. Trav Stat Biol Roscoff (NS) 12:1–356

Elsevier SM, Lipps HJ, Steinbrück G (1978) Histone genes in macronuclear DNA of the ciliate Stylonychia mytilus. Chromosoma (Berl) 69:291–306

Engberg J, Christiansen G, Leick V (1974) Autonomous rDNA molecules containing single copies of the ribosomal RNA genes in the macronucleus of Tetrahymena pyriformis. Biochem Biophys Res Commun 59:1356–1365

Engberg J, Andersson P, Leick V, Collins J (1976) The free rDNA molecules from Tetrahymena pyriformis GL, are giant palindromes. J Mol Biol 104:455–470

Epstein LM, Forney JD (1984) Mendelian and Non-Mendelian mutations affecting surface antigen expression in Paramecium. Mol Cell Biol 4:1583–1590

Fauré-Fremiet E (1954) Réorganization du type endomixique chez les Loxodidae et chez Centrophorella. J Protozool 1:20–27

Fernandez-Galiano D (1979) Transfer of the widely known "Spirotrich" ciliate, Bursaria truncatella O.F. Müller, to the Vestibulifera as a separate order there, the Bursariomorphida. Trans Am Microsc Soc 98:447–454

Findly RC, Gall JG (1978) Free ribosomal RNA genes in Paramecium are tandemly repeated. Proc Natl Acad Sci USA 75:3312–3316

Findly RC, Gall JG (1980) Organization of ribosomal genes in Paramecium tetraurelia. J Cell Biol 84:547–549

Fujishima M, Watanabe T (1981) Transplantation of germ nuclei in Paramecium caudatum. III. Role of germinal micronucleus in vegetative growth. Exp Cell Res 132:47–56

Forney JD, Epstein LM, Preer LB, Rudman BM, Widmayer DJ, Klein WH, Preer JR (1983) Structure and expression of genes for surface proteins in Paramecium. Mol Cell Biol 3:466–474

Gall JG (1974) Free ribosomal RNA genes in the macronucleus of Tetrahymena. Proc Natl Acad Sci USA 71:3078–3081

Gall JG (1981) Chromosome structure and the C-value paradox. J Cell Biol 91:3s–14s

Gall JG, Yao M-C, Blackburn EH, Findly RC, Wild M (1979) The extrachromosomal ribosomal DNA of Tetrahymena and Paramecium. In: Engberg J, Kleenow H, Leick V (eds) Specific eukaryotic genes. Alfred Benzon Symp, vol 13. Munksgaard, Copenhagen, pp 229–305

Gaude H (1981) Zur Funktion der Ciliaten-Riesenchromosomen: Nachweis von RNS-Polymerase in den polytänen Chromosomen der Makronukleusanlage von Stylonychia mytilus. Arch Protistenkd 124:252–258

Gibson I, Martin N (1971) DNA amounts in the nuclei of Paramecium aurelia and Tetrahymena pyriformis. Chromosoma (Berl) 35:374–382

Gillis M, DeLey J, DeCleeve M (1970) The determination of molecular weight of bacterial genome DNA from renaturation rates. Eur J Biochem 12:143–153

Glover CVC, Gorovsky MA (1978) Histone-histone interactions in a lower eukaryote, Tetrahymena. Biochemistry 17:5705–5713

Glover DM, Zaha A, Stocker AJ, Santelli RV, Pueyo MT, DeToledo SM, Lara FJS (1982) Gene amplification in Rhynchosciara salivary gland chromosomes. Proc Natl Acad Sci USA 79:2947–2951

Golikova MN (1965) Der Aufbau des Kernapparates und die Verteilung der Nukleinsäuren und Proteine bei Nyctotherus cordiformis. Arch Protistenkd 108:191–216

Gorovsky MA (1973) Macro- and micronuclei of Tetrahymena pyriformis: a model system for studying the structure and function of eukaryotic nuclei. J Protozool 20:19–25

Gorovsky MA (1980) Genome organization and reorganization in Tetrahymena. Annu Rev Genet 14:203–239

Gorovsky MA, Keevert JB (1975) Subunit structure of a naturally occurring chromatin lacking histones F1 and F3. Proc Natl Acad Sci USA 72:3536–3540

Gorovsky MA, Woodard J (1969) Studies on nuclear structure and function in Tetrahymena pyriformis. I. RNA synthesis in macro- and micronuclei. J Cell Biol 42:673–682

Gorovsky MA, Hattman S, Pleger GL (1973) 6N-methyl adenine in the nuclear DNA of a eukaryote, Tetrahymena pyriformis. J Cell Biol 56:697–701

Gorovsky MA, Keevert JB, Pleger GL (1974) Histone F1 of Tetrahymena macronuclei: unique electrophoretic properties and phosphorylation of F1 in an amitotic nucleus. J Cell Biol 61:134–145

Gorovsky MA, Glover C, Johmann CA, Keevert JB, Mathis DJ, Samuelson M (1978) Histones and chromatin structure in Tetrahymena macro- and micronuclei. Cold Spring Harbor Symp Quant Biol 42:493–503

Grassé P-P (1952) Traité de zoologie, vol 1, fasc I. Masson et Cie, Paris, p 87

Grell KG (1949) Die Entwicklung der Makronucleusanlage im Exkonjuganten von Ephelota gemmipara Hertwig. Biol Zentralbl 68:289–312

Grell KG (1953) Die Konjugation von Ephelota gemmipara Hertwig. Arch Protistenkd 98:287–326

Grell KG (1973) Protozoology. Springer, Berlin Heidelberg New York

Grell KG (1979) Cytogenetic systems and evolution in Foraminifera. J Foram Res 9:1–13

Hadži J (1963) The evolution of the Metazoa. Pergamon, New York

Hamana K, Iwai K (1971) Fractionation and characterization of Tetrahymena histone in comparison with mammalian histones. J Biochem (Tokyo) 69:1097–1111

Hanson ED (1963) Homologies and the ciliate origin of the Eumetazoa. In: Dougherty EC (ed) The lower metazoa. Univ California Press, Berkeley

Helftenbein E (1985) Nucleotide sequence of a macronuclear DNA molecule coding for α-tubulin from the ciliate Stylonychia lemnae. Special codon usage: TAA is not a translation termination codon. Nucleic Acids Res 13:415–433

Hennig W (1966) Phylogenetic systematics. Univ Illinois Press, Urbana

Hill RJ, Stollar BD (1983) Dependence of Z-DNA antibody binding to polytene chromosomes on acid fixation and DNA torsional strain. Nature (Lond) 305:338–340

Hutchinson F (1961) Molecular basis for action of ionizing radiations. Science (Wash DC) 134:533–538

Iwamura Y, Sakai M, Mita T, Muramatsu M (1979) Unequal gene amplification and transcription in the macronucleus of Tetrahymena pyriformis. Biochemistry 18:5289–5294

Iwamura Y, Sakai M, Muramatsu M (1982) Rearrangement of repeated sequences during development of macronucleus in Tetrahymena thermophila. Nucleic Acids Res 10:4279–4291

Jareno MA, Alonso P, Pérez-Silva J (1972) Identification of some puffed regions in the polytene chromosomes of Stylonychia mytilus. Protistologica 8:237–243

Johmann CA, Gorovsky MA (1976a) Purification and characterization of the histones associated with the macronucleus of Tetrahymena. Biochemistry 15:1249–1256

Johmann CA, Gorovsky MA (1976b) An electrophoretic comparison of the histones of various strains of Tetrahymena pyriformis. Arch Biochem Biophys 175:694–699

Kaine BP, Spear BB (1980) Putative actin genes in the macronucleus of Oxytricha fallax. Proc Natl Acad Sci USA 77:5336–5340

Kaine BP, Spear BB (1982) Nucleotide sequence of a macronuclear gene for actin in Oxytricha fallax. Nature (Lond) 295:430–432

Kaney AR, Speare VJ (1983) An amicronucleate mutant of Tetrahymena thermophila. Exp Cell Res 143:461–467

Karrer KM (1983) Germ line-specific DNA sequences are present on all five micronuclear chromosomes in Tetrahymena thermophila. Mol Cell Biol 3:1909–1919

Karrer K, Gall J (1975) The macronuclear ribosomal DNA of Tetrahymena is a palindrome. J Cell Biol 67:202a

Karrer K, Gall J (1976) The macronuclear ribosomal DNA of Tetrahymena is a palindrome. J Mol Biol 104:421–453

Karrer K, Stein-Gavens S, Allitto BA (1984) Micronucleus-specific DNA sequences in an amicronucleate mutant of Tetrahymena. Dev Biol 105:121–129

Katzen AL, Cann GM, Blackburn EH (1981) Sequence-specific fragmentation of macronuclear DNA in a holotrichous ciliate. Cell 24:313–320

Kavenoff R, Zimm BH (1973) Chromosome-sized DNA molecules from Drosophila. Chromosoma (Berl) 41:1–27

Kimmel AR, Gorovsky MA (1976) Numbers of 5S and tRNA genes in macro- and micronuclei of Tetrahymena pyriformis. Chromosoma (Berl) 54:327–337

King BO, Yao M-C (1982) Tandemly repeated hexanucleotide at Tetrahymena rDNA free end is generated from a single copy during development. Cell 31:177–182

Kiss GB, Amin AA, Pearlman RE (1981) Two separate regions of the extrachromosomal ribosomal deoxyribonucleic acid of Tetrahymena thermophila enable autonomous replication of plasmids in Saccharomyces cerevisiae. Mol Cell Biol 1:535–543

Klobutcher LA, Swanton MT, Donini P, Prescott DM (1981) All gene-sized DNA molecules in four species of hypotrichs have the same terminal sequence and an unusual 3'terminus. Proc Natl Acad Sci USA 78:3015–3019

Klobutcher LA, Jahn CL, Prescott DM (1984) Internal sequences are eliminated from genes during macronuclear development in the ciliated protozoan, Oxytricha nova. Cell 36:1045–1055

Kloetzel JA (1970) Compartmentalization of the developing macronucleus following conjugation in Stylonychia and Euplotes. J Cell Biol 47:395–407

Kovaleva VG, Raikov IB (1973) Ultrastructure de l'appareil nucléaire de Trachelonema sulcata Kovaleva, Cilié holotriche gymnostome à macronoyaux diploides. Protistologica 9:471–480

Kovaleva VG, Raikov IB (1978) Diminution and re-synthesis of DNA during development and senescence of the "diploid" macronuclei of the ciliate Trachelonema sulcata (Gymnostomata, Karyorelictida). Chromosoma (Berl) 67:177–192

Lauth MR, Spear BB, Heumann J, Prescott DM (1976) DNA of ciliated protozoa: DNA sequence diminution during macronuclear development of Oxytricha. Cell 7:67–74

Lawn RM, Heumann JM, Herrick G, Prescott DM (1978) The gene-sized DNA molecules of Oxytricha. Cold Spring Harbor Symp Quant Biol 42:483–492

Lee G-S, Lewis SA, Wilde CD, Cowan NJ (1983) Evolutionary history of a multigene family: an expressed human α-tubulin and three processed pseudogenes. Cell 33:477–487

Levine ND, Corliss JO, Cox FEG, Deroux G, Grain J, Honigberg BM, Leedale GF, Loeblich III AR, Lom J, Lynn D, Merinfeld EG, Page FC, Poljansky G, Sprague V, Vavra J, Wallace FG (1980) A newly revised classification of the Protozoa. J Protozool 27:37–58

Lipps HJ (1980) In vitro aggregation of the gene-sized DNA molecules of the ciliate, Stylonychia mytilus. Proc Natl Acad Sci USA 77:4104–4107

Lipps HJ, Erhard P (1981) DNase I hypersensitivity of the terminal inverted repeat DNA sequences in the macronucleus of the ciliate, Stylonychia mytilus. FEBS Lett 126:219–222

Lipps HJ, Hantke KG (1975) Studies on the histones of the ciliate, Stylonychia mytilus. Chromosoma (Berl) 49:309–320

Lipps HJ, Morris NR (1977) Chromatin structure in the nuclei of the ciliate, Stylonychia mytilus. Biochem Biophys Res Commun 74:230–234

Lipps HJ, Steinbrück G (1978) Free genes for rRNAs in the macronuclear genome of the ciliate, Stylonychia mytilus. Chromosoma (Berl) 69:21–26

Lipps HJ, Sapra GR, Ammermann D (1974) The histones of the ciliated protozoan, Stylonychia mytilus. Chromosoma (Berl) 45:273–280

Lipps HJ, Nordheim A, Lafer EM, Ammermann D, Stollar BD, Rich A (1983) Antibodies against Z DNA react with the macronucleus but not the micronucleus of the hypotrichous ciliate, Stylonychia mytilus. Cell 32:435–441

Lipscombe DL, Corliss JO (1982) Stephanopogon, a phylogenetically important "ciliate", shown by ultrastructural studies to be a flagellate. Science (Wash DC) 215:303–304

Lwoff A (1923) Sur la nutrition des infusoires. C R Acad Sci Paris 176:928–930

Mayr E (1963) Animal species and evolution. The Belknap Press of Harvard Univ Press, Cambridge, Mass

McCoy (1976) Updating the tetrahymenids. II. Domestic and natural variation of amicronucleate species of the Tetrahymena pyriformis complex. Acta Protozool 13:235–243

McTavish C, Sommerville J (1980) Macronuclear DNA organization and transcription in Paramecium primaurelia. Chromosoma (Berl) 78:147–164

Merkulova NA, Borchsenius SN (1976) Replication and integration of fragments of new chains into the DNA of the ciliate, Tetrahymena pyriformis GL. Mol Biol (Mosc) 10:1072–1077 (in Russian with English summary)

Meyer GF, Lipps HJ (1980) Chromatin elimination in the hypotrichous ciliate, Stylonychia mytilus. Chromosoma (Berl) 77:285–297

Meyer GF, Lipps HJ (1981) The formation of polytene chromosomes during macronucleus development of the hypotrichous ciliate, Stylonychia mytilus. Chromosoma (Berl) 82:309–314

Meyer GF, Lipps HJ (1984) Electron microscopy of surface spread polytene chromosomes of Drosophila and Stylonychia. Chromosoma (Berl) 89:107–110

Mikami K, Kuhlmann H-W, Heckmann K (1985) Is the initiation of macronuclear DNA synthesis in Euplotes dependent on micronuclear functions? J Protozool in press

Möhn E (1984) System und Phylogenie der Lebewesen, vol 1. Schweizerbart, Stuttgart

Murti KG, Prescott DM (1970) Micronuclear ribonucleic acid in Tetrahymena pyriformis. J Cell Biol 47:460–467

Murti KG, Prescott DM (1983) Replication forms of the gene-sized DNA molecules of hypotrichous ciliates. Mol Cell Biol 3:1562–1566

Nanney DL (1964) Macronuclear differentiation and subnuclear assortment in ciliates. In: Locke M (ed) The role of chromosomes in development. Academic Press, New York, pp 253–273

Nanney DL (1980) Experimental ciliatology. Wiley, New York

Nanney DL, McCoy JW (1976) Characterization of the species of the Tetrahymena pyriformis complex. Trans Am Micros Soc 95:664–682

Nanney DL, Preparata RM (1979) Genetic evidence concerning the structure of the Tetrahymena thermophila macronucleus. J Protozool 26:2–9

Nasmyth KA (1982) Molecular genetics of yeast mating type. Annu Rev Genet 16:439–500

Ng SF, Newman A (1984) The role of the micronucleus in stomatogenesis in sexual reproduction of Paramecium tetraurelia: Micronuclear and stomatogenic events. Protistologica 20:43–64

Nock A (1981) RNA and macronuclear transcription in the ciliate, Stylonychia mytilus. Chromosoma (Berl) 83:209–220

Nouzarède M (1976) Cytology fonctionelle et morphologie expérimentale de quelques Protozoaires ciliés mésopsammique géant de la famille des Geleiidae (Kahl). Bull State Biol Arcachon (Suppl) NS 28:1–315

Ohama T, Kumazaki T, Hori H, Osawa S (1984) Evolution of multicellular animals as deduced from 5S rRNA sequences: a possible early emergence of the Mesozoa. Nucleic Acids Res 12:5101–5108

Oka Y, Honjo T (1983) Common terminal repeats of the macronuclear DNA are absent from the micronuclear DNA in hypotrichous ciliate, Stylonychia pustulata. Nucleic Acids Res 11:4325–4333

Oka Y, Shiota S, Nakai S, Nishida Y, Okulbo S (1980) Inverted terminal repeat sequences in the macronuclear DNA of Stylonychia pustulata. Gene (Amst) 10:301–306

Orgel LE, Crick FHC (1980) Selfish DNA: the ultimate parasite. Nature (Lond) 284:604–607

Orgel LE, Crick FHC, Sapienza C (1980) Selfish DNA. Nature (Lond) 288:645–646

Orias E (1981) Probable somatic DNA rearrangements in mating type determination in Tetrahymena thermophila: a review and a model. Dev Genet 2:185–202

Orias E, Flacks M (1975) Macronuclear genetics of Tetrahymena. I. Random distribution of macronuclear gene copies in Tetrahymena pyriformis syngen 1. Genetics 79:187–206

Pan W-C, Blackburn EH (1981) Single extrachromosomal ribosomal RNA gene copies are synthesized during amplification of the rDNA in Tetrahymena. Cell 23:459–466

Pan WC, Orias E, Flacks M, Blackburn EH (1982) Allele-specific selective amplification of a ribosomal RNA gene in Tetrahymena thermophila. Cell 28:595–604

Pasternak J (1967) Differential genic activity in Paramecium aurelia. J Exp Zool 165:395–417

Pavan C, DaCunha AB (1969) Chromosomal activities in Rhynchosciara and other Sciaridae. Annu Rev Genet 3:425–450

Pelvat B, DeHaller G (1976) Macronuclear DNA in Stentor coeruleus: a first approach to its characterization. Genet Res 27:277–289

Pérez-Silva J, Alonso P (1966) Demonstration of polytene chromosomes in the macronuclear anlage of Oxytrichous ciliates. Arch Protistenkd 109:65–70

Pluta AF, Kaine BP, Spear BB (1982) The terminal organization of macronuclear DNA in Oxytricha fallax. Nucleic Acids Res 10:8145–8154

Pluta AF, Dani GM, Spear BB, Zakian VA (1984) Elaboration of telomers in yeast: recognition and modification of termini from Oxytricha macronuclear DNA. Proc Natl Acad Sci USA 81:1475–1479

Poljansky G (1933/34) Geschlechtsprozesse bei Bursaria truncatella O. F. Müller. Arch Protistenkd 81:420–546

Poljansky G, Sergejeva GI (1981) Autoradiographic investigation of the DNA synthesis during development of the new macronucleus of the ciliate Bursaria truncatella. Tsitologiya 23:666–673 (in Russian with English summary)

Preer JR Jr (1969) Genetics of Protozoa. In: Chen TT (ed) Research in protozoology. Pergamon, Oxford, pp 139–288

Preer JB, Preer LB (1979) The size of macronuclear DNA and its relation to models for maintaining genetic balance. J Protozool 26:14–18

Prescott DM (1983) The C-value paradox and genes in ciliated protozoa. Mod Cell Biol 2:329–352

Prescott DM, Murti KG, Bostock CJ (1973) Genetic apparatus of Stylonychia sp. Nature (Lond) 242:597–600

Prescott DM, Bostock CJ, Murti KG, Lauth MR, Gamow E (1971) DNA of ciliated protozoa. I. Electron microscopic and sedimentation analyses of macronuclear and micronuclear DNA of Stylonychia mytilus. Chromosoma (Berl) 34:355–366

Prescott DM, Heumann JM, Swanton M, Boswell RE (1979) The genome of hypotrichous ciliates. In: Engberg I, Kleenow H, Leick V (eds) Specific eukaryotic genes. Alfred Benzon Symp, vol 13. Munksgaard, Copenhagen, pp 85–99

Proudfoot N (1980) Pseudogenes. Nature (Lond) 286:840–841

Pyne CK (1978) Electron microscopic studies on the macronuclear development in the ciliate, Chilodonella uncinata. Cytobiologie 18:145–160

Pyne CK, Ruch F, Leemann U, Schneider S (1974) Development of the macronuclear anlage in the ciliate Chilodonella uncinata. Chromosoma (Berl) 48:225–238

Radzikowski S (1973) Die Entwicklung des Kernapparates und die Nukleinsäuresynthese während der Konjugation von Chilodonella cucullulus O. F. Müller. Arch Protistenkd 115:419–428

Radzikowski S (1976) DNA and RNA synthesis in the nuclear apparatus of Chilodonella cucullulus O. F. Müller. Acta Protozool 15:47–56

Radzikowski S (1979) Asynchronous replication of polytene chromosome segments of the new macronucleus anlage in Chilodonella cucullulus O. F. Müller. Protistologica 15:521–526

Radzikowski S, Golembiewska M (1977) Chilodonella steini. Morphology and culture method. Protistologica 13:381–389

Rae PMM, Spear BB (1978) Macronuclear DNA of the hypotrichous ciliate Oxytricha fallax. Proc Natl Acad Sci USA 75:4992–4996

Rae PMM, Steele RE (1978) Modified bases in the DNAs of unicellular eukaryotes: an examination of distributions and possible roles, with emphasis on hydroxymethylcytosine. Biosystems 10:37–53

Raikov IB (1959) Der Formwechsel des Kernapparates einiger niederer Ciliaten. II. Die Gattung Loxodes. Arch Protistenkd 104:1–42

Raikov IB (1969) The macronucleus of ciliates. In: Chen T-T (ed) Research in protozoology, vol 3. Pergamon, Oxford, pp 1–128

Raikov IB (1972) Ultrastructure des „capsule nucléaires" („noyaux composés") du Cilié psammophile Kentrophoros latum Raikov 1962. Protistologica 8:299–313

Raikov IB (1976) Evolution of macronuclear organization. Annu Rev Genet 10:413–440

Raikov IB (1982) The protozoan nucleus. Springer, Berlin Heidelberg New York

Raikov IB, Dragesco J (1969) Ultrastructure des noyaux et de quelques organites cytoplasmiques du cilié Tracheloraphis caudatum Dragesco et Raikov (Holotricha, Gymnostomatida). Protistologica 5:193–208

Raikov IB, Morat G (1977) Etude autoradiographique de la synthèse de l'ADN et de l'ARN dans les noyaux du Cilié Loxodes magnus Stokes. Protistologica 13:391–399

Rao MVN, Ammermann D (1970) Polytene chromosomes and nucleic acid metabolism during macronuclear development in Euplotes. Chromosoma (Berl) 29:246–254

Rao MVN, Prescott DM (1967) Micronuclear RNA synthesis in Paramecium caudatum. J Cell Biol 33:281–285

Remane A (1952) Die Besiedlung des Sandbodens im Meere und die Bedeutung der Lebensformtypen für die Ökologie. Verh Dtsch Zool Ges 1951:327–359

Ron A, Urieli S (1977) Qualitative and quantitative studies on DNA and RNA synthesis in Loxodes striatus nuclei. J Protozool 24:150–154

Ruthmann A (1964) Autoradiographische und mikrophotometrische Untersuchungen zur DNA-Synthese im Makronucleus von Bursaria truncatella. Arch Protistenkd 107:117–130

Ruthmann A (1972) Division and formation of the macronuclei of Keronopsis rubra. J Protozool 19:661–666

Schlegel M (1985) Comparative study of allzyme variation in eight species of hypotrichous ciliates (Polyhymenophora, Ciliophora). Zool Syst u Evolut-Forsch 23:171–183

Schwartz V (1958) Chromosomen im Makronucleus von Paramecium bursaria. Biol Zentralbl 77:347–364

Schwartz V, Meister H (1975) Die Extinktion der feulgengefärbten Makronucleusanlage von Paramecium bursaria in der DNA-armen Phase. Arch Protistenkd 117:60–64

Seyffert H-M (1979) Evidence for chromosomal macronuclear substructures in Tetrahymena. J Protozool 26:66–74

Soldo AT, Brickson SA, Lavin F (1981) The kinetic and analytical complexities of the DNA genomes of certain marine and fresh water ciliates. J Protozool 28:377–383

Sommerville J (1970) Serotype expression in Paramecium. Adv Microb Physiol 4:131–178

Sonneborn TM (1957) Breeding systems, reproductive methods and species problems in Protozoa. In: Mayr E (ed) The species problem. Am Assoc Adv Sci Publ, Washington, pp 155–324

Sonneborn TM (1974a) Paramecium aurelia. In: King RC (ed) Handbook of genetics, vol 2. Plenum, New York, pp 469–594

Sonneborn TM (1974b) Tetrahymena pyriformis. In: King RC (ed) Handbook of genetics, vol 2. Plenum, New York, pp 433–467

Sonneborn TM (1975) The Paramecium aurelia complex of 14 sibling species. Trans Am Micros Soc 94:155–178

Spear BB (1980) Isolation and mapping of the rRNA genes in the macronucleus of Oxytricha fallax. Chromosoma (Berl) 77:93–202

Spear BB, Lauth MR (1976) Polytene chromosomes of Oxytricha: biochemical and morphological changes during macronuclear development in a ciliated protozoan. Chromosoma (Berl) 54:1–13

Steinböck O (1963) Origin and affinities of the Lower Metazoa: The "Aceloid" ancestry of the Eumetazoa. In: Dougherty EC (ed) The Lower Metazoa. Univ California Press, Berkeley

Steinbrück G (1976) Untersuchungen zur Organisation des Genomes von Stylonychia mytilus (Ciliata). Thesis Univ Tübingen

Steinbrück G (1983) Overamplification of genes in macronuclei of hypotrichous ciliates. Chromosoma (Berl) 88:156–163

Steinbrück G, Schlegel M (1983) Characterization of two sibling species of the genus Stylonychia (Ciliata, Hypotricha): S. mytilus Ehrenberg, 1838 and S. lemnae n. sp. II. Biochemical characterization. J Protozool 30:294–300

Steinbrück G, Haas I, Hellmer K-H, Ammermann D (1981) Characterization of macronuclear DNA in five species of ciliates. Chromosoma (Berl) 83:199–208

Sugai T, Hiwatashi K (1974) Cytologic and autoradiographic studies of the micronucleus at meiotic prophase in Tetrahymena pyriformis. J Protozool 21:542–548

Swanton MT, Heumann JM, Prescott DM (1980 b) Gene-sized molecules of the macronuclei in three species of hypotrichs: size distribution and absence of nicks. Chromosoma (Berl) 77:217–227

Swanton MT, McCarroll RM, Spear BB (1982) The organization of macronuclear rDNA molecules of four hypotrichous ciliated protozoans. Chromosoma (Berl) 85:1–9

Szostak JW, Blackburn EH (1982) Cloning yeast telomeres on linear plasmid vectors. Cell 29:245–255

Tait A (1970) Enzyme variation between syngens in Paramecium aurelia. Biochem Genet 4:461–470

Toennesen T, Engberg J, Leick V (1976) Studies on the amount and the localization of the tRNA and 5-S rRNA genes in Tetrahymena pyriformis GL. Eur J Biochem 63:399–407

Tonegawa S (1983) Somatic generation of antibody diversity. Nature (Lond) 302:575–581

Torch R (1964) Autoradiographic studies of nucleic acid synthesis in a gymnostome ciliate, Tracheloraphis sp. J Cell Biol 23:98 a

Truett MA, Gall JG (1977) The replication of ribosomal DNA in the macronucleus of Tetrahymena. Chromosoma (Berl) 64:295–303

Vorob'ev VI, Borchsenius SM, Belozerskaya NA, Merkulova NA, Irlina IS (1975) DNA replication in macronuclei of Tetrahymena pyriformis GL. Exp Cell Res 93:253–260

Wells C (1960) The response of Tetrahymena pyriformis to ionizing radiation: strain specific radiosensitivities. J Cell Comp Physiol 55:207–219

Werz G (1974) Fine structural aspects of morphogenesis in Acetabularia. Int Rev Cytol 30:319–367

Wesley RD (1975) Inverted repetitious sequences in the macronuclear DNA of hypotrichous ciliates. Proc Natl Acad Sci USA 72:678–682

Williams JB, Fleck EW, Hellier LE, Uhlenhopp E (1978) Viscoelastic studies on Tetrahymena macronuclear DNA. Proc Natl Acad Sci USA 75:5062–5065

White T, Allen S (1984) Eliminated sequences and a specific site of methylation in the macronucleus of Tetrahymena thermophila. In: Abstracts of papers presented at the meeting on "Ciliate molecular genetics" 2.5.–6.5.1984. Cold Spring Harbor Laboratory, New York, p 11

Wild MA, Gall JG (1979) An intervening sequence in the gene coding for 25S ribosomal RNA of Tetrahymena pigmentosa. Cell 16:565–573

Wollgiehn R (1982) Control of morphogenesis in Acetabularia. In: Nover L, Luckner M, Parthier B (eds) Cell differentiation. Springer, Berlin Heidelberg New York, pp 529–548

Woodard J, Kaneshiro E, Gorovsky MA (1972) Cytochemical studies on the problem of macronuclear subnuclei in Tetrahymena. Genetics 70:251–260

Wünning IU, Lipps HJ (1983) A transformation system for the hypotrichous ciliate, Stylonychia mytilus. EMBO J 2:1753–1757

Yao M-C (1981) Ribosomal RNA gene amplification in Tetrahymena may be associated with chromosome breakage and DNA elimination. Cell 24:765–774

Yao M-C (1982 a) Elimination of specific DNA sequences from the somatic nucleus of the ciliate, Tetrahymena. J Cell Biol 92:783–789

Yao M-C (1982b) Amplification of ribosomal RNA genes in Tetrahymena. In: Busch H, Rothblum L (eds) The cell nucleus. Academic, New York, pp 127–153

Yao M-C, Gall JG (1977) A single integrated gene for ribosomal RNA in a eucaryote, Tetrahymena pyriformis. Cell 12:121–132

Yao M-C, Gorovsky MA (1974) Comparison of the sequences of macro- and micronuclear DNA of Tetrahymena pyriformis. Chromosoma (Berl) 48:1–18

Yao M-C, Yao CH (1981) The repeaed hexanucleotide C-C-C-C-A-A is present near free ends of macronuclear DNA of Tetrahymena. Proc Natl Acad Sci USA 78:7436–7439

Yao M-C, Blackburn E, Gall JG (1979) Amplification of rRNA genes in Tetrahymena. Cold Spring Harbor Symp Quant Biol 43:1293–1296

Yao M-C, Blackburn E, Gall JG (1981) Tandemly repeated C-C-C-C-A-A hexanucleotide of Tetrahymena rDNA is present elsewhere in the genome and may be related to the alteration of the somatic genome. J Cell Biol 90:515–520

Yao M-C, Kimmel AR, Gorovsky MA (1974) A small number of cistrons for ribosomal RNA in the germinal nucleus of a eukaryote, Tetrahymena pyriformis. Proc Natl Acad Sci USA 71:3082–3086

Yao M-C, Choi J, Yokoyama S, Austerberry CF, Yao C-H (1984) DNA elimination in Tetrahymena: a developmental process involving extensive breakage and rejoining of DNA at defined sites. Cell 36:433–440

Yokoyama RW, Yao M-C (1982) Elimination of DNA sequences during macronuclear differentiation in Tetrahymena thermophila as detected by in situ hybridization. Chromosoma (Berl) 85:11–22

Zech L (1964) Zytochemische Messungen an den Zellkernen der Foraminiferen, Patellina corrugata and Rotaliella heterokaryotica. Arch Protistenkd 107:295–330

Zier H (1972) Cytologische Untersuchungen zur Konjugation von Euplotes aediculatus. Thesis Univ Münster

Heterochromatin and Germ Line-Restricted DNA

W. HENNIG [1]

The three reviews assembled in this volume approach the question of germ line-soma differentiation from the viewpoint of three different organisms. In all three organisms germ line and somatic cells are substantially distinguished in their karyotypes. Entire chromosomes or sections thereof are excluded from the somatic cells. This is achieved by quite different mechanisms. The divergent development of the karyotypes occurs within the germ line (as, for example, in *Sciara*) or during the earliest development of the organism (as, for example, in *Ascaris*). In ciliates an even more complex sequence of DNA rearrangements is found.

The examples for germ line-soma differentiation at the DNA level are by no means exhaustive. However, considerable attention has been given to these systems during the last years. For other organisms with different types of germ line-soma differentiation reviews of earlier work are available. They are not considered in more detail, since little progress has been made recently. Such reviews concern, for example, B chromosomes (Jones, 1977; Müntzing 1974), and the complicated processes of differential heterochromatization in coccids (Hughes-Schrader 1948; Brown and Nur 1964).

Other kinds of genomic rearrangement during development, such as in the immune system (see Nover et al. 1982), the amplification of ribosomal DNA (see Beckingham 1982), or the chorion genes during oogenesis (see Spradling and Rubin 1981), as the amplification of DNA sequences coding for certain enzymes under selective pressure (Schimke 1982; Stark and Wahl 1984; Hamlin et al. 1984) are also examples of differentiation at the DNA level. These cases may, however, be considered as more specialized phenomena.

From the reviews in this book it is evident that fundamental questions remain without answer: What is the biological reason for eliminating parts of the genome from somatic cells? Is there any biological function connected with the germ line-limited parts of the genome? If so, what is the nature of this function?

One of the most obvious characteristics of eliminated chromatin is its heterochromatic nature. Any assessment of the biological role of germ line-restricted parts of the genome leads, therefore, inevitably to another major open question concerning genome structure in eukaryotes, the question of the nature and biological function of heterochromatin. In this chapter, some considerations of this question are presented.

---

[1] Genetisch Laboratorium, Katholieke Universiteit, 6525 ED Nijmegen, The Netherlands.

Results and Problems in Cell Differentiation 13
Germ Line – Soma Differentiation (Ed. by W. Hennig)
© Springer-Verlag Berlin Heidelberg 1986

1 What is Heterochromatin?

Since the early days of cytology it has been known that some chromosomes
or parts thereof retain a strong staining behavior during the interphase, while the
rest of the genome is decondensed. This phenomenon was called heteropyknosis
(Gutherz 1907). Heitz (1928, 1929) introduced the term heterochromatin for
chromosomes or chromosome regions which differ in their staining behavior during the interphase due to their relatively condensed state. We now know that this
intense staining is a consequence of an increased concentration of DNA caused
by more extensive condensation than in the rest of the genome. Strongly staining
chromosome regions are characteristically associated with the kinetochore regions ("centric heterochromatin"), with telomeres, with nucleolus organizer regions (NOR's), and with sex chromosomes. Entire chromosome arms or even
whole chromosomes may also be heterochromatic (for examples see the chapters
by Tobler and Gerbi, this Vol.). Another, and probably the only further common
character of heterochromatin is its asynchronous, usually late replication (Lima-
de-Faria 1959), which could be a consequence of the structural organization of
these chromosome regions at a level above the nucleosomal organization.

The discovery that some parts of the genome remain in a condensed state during interphase has induced cytologists to search for an explanation of this phenomenon. A most plausible explanation is that such chromosome regions are inert, i.e., not transcriptionally active, so that the DNA remains tightly packed.
This idea is, in particular, reflected in Brown's distinction (1966) of "facultative"
and "constitutive" heterochromatin. Brown (1966) defined as facultative heterochromatic those condensed chromosomes or chromosome regions which find
no equivalent in their homologous chromosome. This applies especially to sex
chromosomal heterochromatin. Chromosomes or chromosome regions being
heterochromatic in both homologues, for example kinetochore-associated heterochromatin, were characterized as constitutive heterochromatin.

This distinction of facultative and constitutive heterochromatin was based on
the notion that facultative heterochromatin was only heterochromatic under certain developmental conditions and that its condensed state therefore reflected a
transiently inactive state of the respective genetic material. Well-known examples
of facultative heterochromatin are the mammalian X chromosome (review Lyon
1968) and the peculiar situation in some coccid insects, where the paternal haploid
genome in offspring males becomes heterochromatic and functionally inactive
(Brown and Nur 1964).

Although initially not intended, this terminology has for many authors implied that "constitutively" heterochromatic chromosome regions are invariably
condensed and inactive – an idea which can be found even in some modern text
books of genetics. This, however, is incorrect, as can be easily seen from early cy-
togenetic-studies of heterochromatin (for example Cooper 1959). This point will
be discussed in some detail below.

Extensive biochemical studies were directed towards an understanding of the
molecular properties of heterochromatin. It seemed reasonable to assume that facultative heterochromatin in its molecular structure was basically comparable to
euchromatin because of its transient heterochromatic character. Although some

cytologists postulated that constitutive heterochromatin was structurally, and in its molecular composition, basically different from euchromatin, others, including Brown (1966), explicitly denied such a difference. The concept of molecular differences of constitutive heterochromatin was largely induced by the discovery of the relative genetic inertness of such heterochromatic chromosome sections (see Schultz 1947).

After the discovery of repetitive DNA sequences as characteristic components of the eukaryotic genome by Britten and his associates (Britten and Kohne 1968), it was soon recognized that highly repetitive DNA sequences (often described as satellite DNA) are major constituents of constitutively heterochromatic chromosome regions (reviews: Walker 1971; Flamm 1972; Rae 1972; Bostock 1971; Hennig 1973; John and Micklos 1979). This was consistent with the results from the genetic analyses implying a relative genetic inertness of heterochromatin, since highly repetitive DNA sequences appeared to be rarely, if at all, transcribed (Flamm 1972), and their nucleotide sequence is often not compatible with protein coding functions (Southern 1970; Flamm 1972; see also Singer 1982; John and Micklos 1982).

On the basis of this knowledge in particular, there has been a strong tendency to consider heterochromatin as "junk" or "dispensive" material which has accumulated during evolution (Ohno 1973). This view has, however, been opposed by others suspecting chromosome "housekeeping" functions or other unknown biological functions related with heterochromatin (Hennig et al. 1970; see discussion in Ohno 1973, pp 188–190; Walker 1971; Rae 1972; Bostock 1971; Hilliker and Appels 1980; John and Micklos 1971; see also Lima-de-Faria 1983).

During the past few years, little essentially new information on the molecular and functional properties of heterochromatin has been obtained. It was recognized (Hennig et al. 1970; Kram et al. 1972; Brutlag 1980) that the DNA sequence composition of constitutive heterochromatin is more complex than initially assumed. In some instances transcripts were discovered (Varley et al. 1980; Wu et al. 1986). New insight to the molecular structure of heterochromatin emerged from our study of the Y chromosome of *Drosophila hydei* (see Hennig 1985). Some of the fundamental features of the DNA composition of this chromosome will be discussed later.

2 Heterochromatic Appearance Reflects Functional Inactivity

It is well known from classic cytology that heterochromatic chromosomes or chromosome regions can be allocyclic. Thus, they may not only appear positively heteropyknotic, which means more condensed than the rest of the genome, but also negatively heteropyknotic. Such a negative heteropyknosis can be experimentally induced, for example by temperature (cold treatment) (Wolff 1957).

Another example of allocyclic behavior of heterochromatin, which is related to the developmental state of a cell, is the oogenesis of the copepod *Cyclops* (Beermann 1959, 1977). Somatic chromosomes of some species of *Cyclops* are devoid of heterochromatin. However, in most germ cell stages, major sections of the *Cyclops* chromosomes are heterochromatic. This difference in the karyotypes of somatic and germ line cells is accounted for by chromatin elimination between the 5th and the 7th cleavage division in early development, comparable to the diminution of DNA in *Ascaris* chromosomes (see Tobler, this Vol.). However, the sections of the *Cyclops* chromosomes, which are subjected to a subsequent elimination from the genome, in the earliest development do not display this property of being heterochromatic except after the first few cleavages have been passed. Only then is a heterochromatic state achieved. During the 5th to 7th cleavage division these heterochromatic chromosome sections are eliminated from those cells developing into somatic cells. Absence of heterochromatin can also be observed during oogenesis. During the leptotene all chromosomes display a lampbrush-like structure with few indications for the presence of heterochromatin, except for some centric heterochromatin. It is, therefore, obvious that the heterochromatic sections of these chromosomes are not permanently condensed, although they seem in other respects to fit the characters of constitutive heterochromatin.

These observations certify that even constitutive heterochromatin may occur in two alternative states, a condensed and a decondensed state. It is difficult to say whether any heterochromatin exists which resides permanently in a condensed state. Probably there is no multicellular organism in which the cytology of chromosomes has been studied throughout all stages of development. In particular, studies on early embryogenesis are almost absent. The most extensively studied chromosomes of animals are those of *Drosophila*, but even here the investigations have been restricted to a small number of tissues and developmental stages. In mitotic metaphases of *D. melanogaster*, mainly from neuroblasts, major blocks of heterochromatin are associated with the kinetochore regions of the autosomes and with the sex chromosomes. In the male germ line, however, heterochromatin is almost absent. I shall describe this situation in some detail for *Drosophila hydei* (Kremer, Hennig and Dijkhof 1986).

In the genome of *Drosophila hydei*, the entire Y chromosome and one arm of the metacentric X chromosome are heterochromatic if somatic metaphase chromosomes are studied. Additional heterochromatin is found in the kinetochores of the autosomes. According to conventional nomenclature, these chromosome regions would be considered as constitutively heterochromatic. In particular, the X heterochromatin and the kinetochore-associated heterochromatin of the autosomes fullfil the criterion of being rich in highly repetitive DNA (Hennig et al. 1970; Renkawitz 1978 b; Hennig 1972). This is less true for the Y chromosome in *D. hydei*, which has been demonstrated to carry relatively small proportions of satellite-like DNA (Hennig 1972; Renkawitz 1978a; see Hennig 1985). Moreover, the Y chromosome may not be entirely heterochromatic in neuroblast interphases, since Y chromosomal DNA sequences are not entirely restricted to the heterochromatic blocks found in these cells (cf. Hennig et al. 1983). This may raise doubts as to the character of heterochromatin (e.g., constitutive vs. facultative) of the Y chromosome. Also in other *Drosophila* species and in other organisms the actual designation of Y chromosomes as constitutive heterochromatin is not used in a consistent way (cf. Beermann 1962, pp 118).

The chromosomal cytology during spermatogenesis is clearly not compatible with a permanently heterochromatic character of either the Y chromosomal or the X chromosomal heterochromatin. It has long been known that the Y chromosome becomes decondensed in the primary spermatocytes in *Drosophila* (Meyer et al. 1961) and remains active in RNA synthesis during this entire developmental stage (Hennig 1967). There is no cytological evidence for a residual portion of heterochromatin in these nuclei, as we have recently demonstrated, although minor blocks may remain undetected. The Y as well as the X chromosome are at least in major parts allocyclic (Kremer et al. 1986). Both chromosomes condense prior to the autosomes at the onset of the meiotic metaphase. The heterochromatic arm of the X chromosome, which, according to the classic definition, is considered as constitutively heterochromatic, is negatively heteropyknotic during the meiotic metaphases.

This observation clearly raises the question whether constitutively heterochromatic chromosome regions in other organisms may also eventually become decondensed. I doubt whether this possibility can at present be excluded. For example, the failure of *Cyclops* meiotic and early cleavage chromosomes to display heterochromatin has already been described. Also during the early cleavages in *Drosophila* development no cytological distinction between eu- and heterochromatin is possible (Sonnenblick 1950). Similar observations were made for early mammalian development (see Rae 1972). It is therefore not unlikely that all constitutive heterochromatin becomes decondensed at certain developmental stages, as has already been implied by Cooper (1959).

Another intriguing example in support of this view is provided by the heterochromatic sections in the genome of *Microtus agrestis*, which are typically constitutively heterochromatic (Matthey 1949). In thymus cells cultured in vitro, the heterochromatic sections of the chromosomes can be induced to decondense after treatment with thyroid hormones (Schneider et al. 1973). It is thus possible to induce typically constitutively heterochromatic chromosome sections into a decondensed, and hence probably functional, state by agents which are typically considered as regulatory for transcription. A somewhat related situation may exist in the case of *Spiranthes sinensis*, described by Tanaka (1969a, b cit. from D'Amato 1977). The karyotype of this orchid consists of 2n = 30 chromosomes. Only few of these 30 chromosomes are entirely euchromatic, the rest being to a great extent heterochromatic. In the context of our discussion it is of interest that the number of euchromatic chromosomes varies according to the topological origin and physiology of the plant. In cold-moor populations karyotypes with only two euchromatic chromosomes predominate while in warm-moor and grassy land populations the karyotype contains up to six entirely euchromatic chromosomes. The ecological differences are accompanied by morphological and presumably physiological differences such as the larger size of the plants with more euchromatic chromosomes. This obvious relationship between the eu- and heterochromatic state of some of the chromosomes and the environmental conditions suggests the location of genetic information with relevance for the physiological state of the organism in heterochromatic regions of the genome.

In summary the data discussed in this section clearly reveal variable states of so-called constitutive heterochromatin, and thus suggest that the suppositions usually ascribed to constitutive heterochromatin probably are inadequate to describe structure and function of this portion of the genome.

3 Heterochromatin Is Not Exclusively Composed of Highly Repetitive DNA

An important step towards the exploration of the molecular structure of heterochromatin was made after the advent of in situ hybridization. Some of the first conclusions reached with this technique was that highly repetitive DNA sequences are enriched and preferentially located in constitutive heterochromatin (cf. Hennig 1973). This finding, together with earlier results of studies on highly repetitive DNA, led to the following view.

Constitutive heterochromatin is to a large extent composed of highly repetitive DNA. The nucleotide sequences of the DNA repeats are species-specific and so are the locations of highly repetitive DNA sequences. Different blocks of heterochromatin may be composed of different types of highly repetitive DNA (often seen as satellites in Cs-salt equilibrium centrifugation). In some instances considerable nucleotide sequence divergence has been found within a distinct satellite DNA fraction (for example in *M. musculus*: Southern 1975), while in other – however, more exceptional – cases the nucleotide sequences are highly homogeneous (for example, *D. virilis*: Gall and Atherton 1974). Highly repetitive DNA fractions originated in evolution from a short sequence (sometimes not more than six nucleotides long), which by amplification became tandemly repeated many times, usually in several separated sequence clusters. Large-scale repeat patterns are superimposed on the primary repeat patterns. This implies that the evolutionary pathway of these blocks of highly repetitive DNA sequences involved several steps of multiplication. Each step has included variable numbers of (usually diverged) copies of the basic sequence (Southern 1975; Walker 1971). For satellite DNA sequences which had been studied in more detail, no transcripts had been discovered (Flamm 1972; Walker 1971) and the nucleotide sequence itself argues against a protein coding role (John and Micklos 1979).

This general concept, however, has subsequently been modified and refined in several respects. Several authors demonstrated that other repetitive or single copy sequences are interspersed between the satellite sequences (Brutlag 1980; Kram et al. 1972). Also the species specificity of satellite DNA is not as strict as originally claimed, although its evolutionary conservation is more restricted than for other sequences. The studies of Varley et al. (1980) and Baldwin and Macgregor (1985) on lampbrush chromosomes of various newts demonstrated the presence of copies of highly repetitive DNA sequences in transcripts. Also in oocytes of the newt *Notophthalmus*, satellite-like (i.e., short, tandemly repeated) DNA sequences are cotranscribed with the closely linked histone genes (Diaz and Gall 1985). These results clearly necessitate a reconsideration of the question of transcriptive activity of such satellite DNA sequences.

Already in the early studies it had been shown that satellite-related DNA sequences were not restricted to large blocks containing tandem repeats of a distinct sequence, but that some copies were spread all over the genome (Hennig et al. 1970). This has recently been confirmed by DNA sequencing. In the region of the *white* locus in the X chromosome of *D. melanogaster*, for example, a single copy of the basic repeat of the $1.68 \times g \times cm^{-3}$ satellite was discovered (Tartof et al.

1984). This clearly shows that satellite DNA sequence copies are not restricted to heterochromatin. Single copies or small blocks of satellite DNA sequences occur scattered throughout the genome. Heterochromatin may be characterized by containing larger, and therefore cytologically visible, blocks of such sequences which may otherwise constitute normal components of the genome in interspersed distribution.

4 The Y Chromosome as Model for Heterochromatin Structure

To consider possible biological functions of heterochromatin, the relationship between sex chromosomes and heterochromatin must first be discussed in more detail. I shall concentrate in this respect to the Y chromosome, assuming it to be representative for the molecular structure of sex chromosomes as far as discussed here.

It has been pointed out earlier that the properties of the Y chromosome do not unequivocally fit to an assignment to constitutive heterochromatin. This can, in more detail, be exemplified on the basis of our recent studies on the molecular structure of the Y chromosome of *Drosophila hydei* (Vogt and Hennig 1983; Vogt et al. 1982; Hennig et al. 1983; Vogt and Hennig 1986a,b; Vogt et al. 1986; Huijser and Hennig 1986; Brand et al., in preparation). They can be summarized in the following conclusions:

1. So far only repetitive DNA sequences have been discovered. These sequences belong to the moderately or middle repetitive sequence class.

2. Two types of sequence must be distinguished (Vogt and Hennig 1983; Hennig et al. 1983): The first type is *Y-specific*, i.e., only present in the Y chromosome. (The presence of single copies in interspersed positions elsewhere in the genome cannot be excluded.) This first sequence type represents relatively short (ca. 200 bp long), tandemly repeated but diverged nucleotide sequences. The copy numbers range up to 1000 copies per genome. The second type (*Y-associated*) of Y chromosomal DNA sequences is not restricted in its location to the Y chromosome, but occurs in several other locations, including the X chromosome. These sequences are often long (several kb) compared to the Y-specific basic sequence. Their copy numbers range between 10 and 100.

3. *Y-specific* DNA sequences are rather species-specific and are not found in more distant *Drosophila* species. The *Y-associated* type is more conserved and occurs not only in other *Drosophila* species, but homologies have also been discovered to mammalian DNA, including man.

4. Both types of DNA sequence occur interspersed in the Y chromosome and are transcribed together into giant transcripts of a length up to 1500 kb (Grond et al. 1983; deLoos et al. 1984). Most of these transcripts are testis-specific and are synthezised exclusively in primary spermatocyte nuclei.

5. Processing of the large primary transcripts leads to the formation of a heterogeneous high molecular weight fraction of RNA molecules which may be restricted to the nucleus (see Lifschytz et al. 1983) and to RNA fractions of discrete sizes which are found in cytoplasmic poly(A)$^+$ RNA (Brand, Hennig, Huijser,

Vogt, unpublished data). Cytoplasmic RNA fractions are derived from the *Y-as-sociated* DNA class.

6. At least some of the non-Y chromosomal copies of the Y-associated DNA sequences are also transcribed in testes and possibly in other tissues. While in their Y chromosomal location the formation of large lampbrush loops accompanies transcription, comparable structures are not found for non-Y-chromosomal loci.

These observations on the Y chromosomal DNA of *Drosophila hydei* may be representative for Y chromosomes in general. All studies on Y chromosomal DNA sequences in other organisms, though much less extensive, are compatible with the pattern discovered for the *Drosophila* chromosome (see Hennig 1986).

The features described for the Y chromosomal DNA of *D. hydei* fit in many aspects to the character of DNA sequences characteristic for constitutive heterochromatin. In particular, the occurrence of satellite-like DNA sequences, arranged in blocks of tandem repeats and interspersed with other sequences is just such a common feature, although the tandem-repeat blocks in the Y chromosome are much smaller than for more typical satellite DNA's. One should, however, in this respect recall that intercalary heterochromatin, which in many aspects is also comparable to constitutive heterochromatin, seems to be composed of small blocks of satellite-like sequences (Hennig et al. 1970). A similar situation must probably be ascribed to the small blocks of interspersed heterochromatin in *Cyclops strenuus* (Beermann 1977).

These considerations suggest that DNA sequences, usually described as satellite DNA fractions, may represent just one extreme of a continuum of sequence arrangements of this type. Similar structural patterns of short, tandemly repeated DNA sequences may occur with all possible frequencies of repeats down to a few repeat copies. The numerous small satellite DNA fractions discovered in calf DNA may be taken as a situation intermediate between extremely high and low copy numbers of satellite DNA sequences. One might speculate that transcription of this type of sequence is simply a consequence of their interspersion with other transcribed sequences. Such an argument may, however, be questioned on the basis of our findings on the structure and expression of Y chromosomal DNA sequences (see below).

5 Are There Possible Germ Line-Related Functions of Heterochromatin?

5.1 Protein Coding and Protein Binding

The function of genes located on the Y chromosome during the meiotic prophase may have no relationship to their location in a heterochromatic environment. Considering other evidence on the cytological state of heterochromatin during the first meiotic prophase, a direct relationship of at least some of these genes seems, however, more likely.

Evidence for Function of Heterochromatin the Germ Line. Let us first consider the evidence for a connection between germ line differentiation and possible functions of heterochromatin connected to this. The strongest argument for *germ line-restricted* functions of, at least, a part of heterochromatin is derived from experimental evidence as assembled in this volume. Alternatively, one would have to assume that the elimination from the germ line is dictated by incompatibility of this portion of the genome with somatic cell functions. A second, although somewhat more indirect argument, for a germ line-restricted function of heterochromatin is the underreplication of heterochromatic chromosome regions, in particular of the sex chromosomes, in insects during the formation of polytene chromosomes. Heterochromatin often remains at a diploid level (Rudkin 1969; Hammond and Laird 1985) in polytene cells. This implies that it is, at least in these cells, not essential. Consistent with this, the situation in the macronuclei (see Steinbrück, this Vol.) suggests that vegetative cells of ciliates have no requirements for highly repetitive DNA.

In the context of our discussion also the observations of Ribbert (unpublished, and Ribbert 1975; see also Hennig 1980), are relevant, who noticed differences between the banding patterns of polytene cells from the trichogene cells and the nurse cells of *Calliphora*. This observation seemed not consistent with observations made in other organisms (Beermann 1952), where a high degree of constancy of the banding patterns of polytene chromosomes with different cell types of one organism was established. Further investigation showed, however, that these general conclusions of Beermann are not affected by the rather special situation in *Calliphora*. The difference in the banding patterns between polytene chromosomes of nurse cells (which are germ line-derived) and trichogene cells (which are somatic cells) proved to be caused by underreplication of satellite DNA's in trichogene cells but not in nurse cells. Since these satellite DNA sequences occur interspersed throughout the genome, nurse cell polytene chromosomes display a banding pattern differing from the one of the polytene chromosomes in trichogene cells by additional bands containing satellite DNA. Thus, here again, DNA sequences typically enriched in heterochromatic portions of the genome are specifically retained in germ line-related cell types.

Another line of evidence for a possible functional state of heterochromatin in germ cells has been mentioned earlier. Often heterochromatic chromosome regions appear negatively heteropyknotic before or during meiotic metaphase. If decondensation of chromosomal material is accepted as indicative for a functional state of the DNA, then some kind of activity of the negatively heteropyknotic chromosome regions must be considered.

Possible Function of Heterochromatin in the Germ Line. If there is any biological role of heterochromatin in the germ line, we have next to ask what kind of function this might be. Most likely it should be directly related to the specific properties of germ cells. One essential biological function of germ cells in both males and females is the differentiation of cells towards the morphology of a gamete. The basic requirements of such morphogenetic pathways are certainly not different from any other cellular differentiation process, even though the degree of morphological complexity may be considered as extremely high. However, at the chromosomal level there are requirements which are not relevant in any other cell type of the organisms. First, the number of chromosomes has to be reduced to a haploid complement. Second, this process of reduction of the chromosome number is preceded by recombination between the chromosomes, an event which requires precise pairing of the chromosomes and other highly sophisticated molecular mechanisms. Pairing of homologues and recombination, although with a lower frequency, can, however, also be observed in somatic cells. There is, nevertheless, one other event in germ cell development, which is not often clearly enough considered. The onset of development of a new organism in the early development in all probability requires a "zero" position of all regulatory signals at the chromosomal level or at least a fundamental revision of signals at the chromosomal level. Such a de-regulation of the genome would most likely occur during germ cell differentiation.

Considering the molecular mechanisms which might be involved in a de-regulation of the genome, it seems reasonable to assume that certain chromosomal proteins must be replaced or entirely removed. How can this be achieved and at what time of the germ cell development may this take place? In general, a suitable time during the cell cycle for reprogramming the cell has been considered to be DNA replication. However, the last premeiotic replication occurs at a time where a major part of the genetic information required for germ cell differentiation is still to be transcribed. A re-programming of the genome in a germ cell would therefore be likely to occur later. Molecular mechanisms other than those taking place in connection with DNA replication may be required. Are there processes in germ cell development other than DNA replication which could account for re-programming events?

For males it is well established that during the postmeiotic differentiation of the sperm a substitution of histones by other basic chromosomal proteins takes place. Moreover, most likely the nucleosomal conformation in the chromatin may, at least in some organisms, be given up in favor of a more dense packaging of the DNA in the head of the male gamete. It is usually argued that this histone substitution is required for a tight packaging of the genome in the sperm head. But obviously, this process would also provide sufficient opportunity to revise regulatory signals at the DNA.

A comparable process does not obviously take place in female gamete development. Although there is little information on the equipment with chromosomal proteins in the late oocyte chromatin, a substantial replacement of histones is not likely and the nucleosomal structure appears to be maintained, at least as long as transcription takes place (Trendelenburg and McKinnell 1979).

Oocyte development, nevertheless, also displays special features, which might be worth considering in the view of our question on a possible de-regulation of the genome. During leptotene and diplotene in many organisms lampbrush chromosomes are formed. It has long been assumed that the transcription in these chromosomes was required to supply the information required during early embryonic development, where no RNA is synthesized. However, objections to such a function of lampbrush chromosomes have been raised. First, the amount and quality of information actually supplied by the maternal genome for early developmental processes is still an open problem, except that there is no doubt that the maternal message can promote early development for quite a long period. Second, it has definitely been established that most of the mRNA species found in the early embryo were already present in the oocyte before the onset of the lampbrush stage (see Davidson 1976). Loop formation may, therefore, have functions other than supplying messengers for the early development. One of such possible functions might be related to reprogramming the maternal genome for early development.

Considering this possibility in more detail, it seems relevant to recall some peculiarities observed in lampbrush chromosomes which are related to transcription. One of these unusual features is the extent of transcription of the genome in these chromosomes. From molecular studies it has been derived that oocytes in the lampbrush stage transcribe a particularly large diversity of genomic sequences (see Davidson 1976). From the view point of cytology, it is clear that transcription is not restricted to the lampbrush loops but occurs widely also along the chromosome axis. Moreover, it has been established that also satellite DNA-like DNA sequences which are usually not expressed (see p. 180) during the lampbrush chromosome stage are represented in transcripts. Diaz et al. (1986) have shown that in *Notophthalmus* histone genes are expressed together with adjacent satellite DNA-like spacer sequences which are probably transcribed over irregular distances without precise termination. This situation is very reminiscent of our observations on the transcription of repeated, satellite DNA-like Y chromosomal DNA sequences in *Drosophila hydei* (see Vogt et al. 1982; Hennig 1985). In fact, so far we have not been able to recover any Y chromosomal DNA sequence not represented in transcripts within testis RNA. Thus, DNA sequences obviously located in heterochromatic chromosome regions seem to be often expressed during germ cell development. It was mentioned before that also during the oocyte development of *Cyclops* a lampbrush chromosome-like stage is passed, which suggests transcription of the large heterochromatic sections of the *Cyclops* chromosomes during germ cell development.

These observations might be indicative of different molecular processes. One simple possibility is that during the premeiotic phase a general de-regulation of the genome occurs, which subsequently causes an overall transcription of the genome. As an alternative, those genomic regions activated in the meiotic prophase – as, for example, heterochromatic chromosome regions – may have special functions at this developmental stage.

From our observations on spermatocyte development in *D. hydei*, we arrived at the conclusion that Y chromosomal genes may have a rather special function (for review see Hennig 1985). We have evidence that one of the fundamental tasks of the Y chromosomal lampbrush loop in *Drosophila* is to bind specific proteins. These proteins might be involved in the reorganization of the chromatin during or after meiosis. One might, moreover, speculate that the formation of lampbrush loops, which clearly is a germ line-specific event, plays a central role in the reorganization and replacement of chromosomal proteins (see Kremer et al. 1986).

Any evaluation of regulatory processes during germ cell differentiation has to account for the phenomenon of imprinting. Imprinting has, in the past, been considered a rather specialized phenomenon. However, the recent observations of Solter and McGrath (1985) that both parental genomes, at least in mammals, appear necessary for a normal early development in mice raises the possibility that imprinting is a rather general phenomenon. The insertion of regulatory signals, specific for the paternal and maternal origin of the genome, would most likely take place during advanced stages of the germ cell differentiation. The occurrence of imprinting would hence not be in contradiction to a possible general reprogramming of the genome during germ cell development.

In conclusion, the experimental data do not exclude, the possibility that heterochromatin has special germ line-related function. It is quite possible that it is involved in a functional reorganization of the genome. To understand the function of heterochromatin it might therefore be essential to understand more on the function of chromosomal proteins. It is clear that the coming years will provide us with more insight into the biological role of chromosomal proteins.

5.2 Has Heterochromatin a Target Function for Transposable Elements?

The complexity of the genome provides ample opportunity for different biological functions apart from protein structure to be encoded. The genome has not only to account for ontogenesis of an organism, but it has also facilitated its evolution and it will facilitate further evolution. The structure of the genome must therefore also be considered from the viewpoint of its development in evolution. From our observations on the fine structure of Y chromosomal DNA in *D. hydei*, we have considered that heterochromation may amplify the flexibility of the genome in the context of evolutionary changes (Hennig 1985). As has been mentioned before, the different copies of Y-associated DNA sequences, which in homologous copies are also present elsewhere in the genome, are rather homogeneous in their nucleotide sequences. This, together with the species-specificity of their genomic locations, recalls the properties of transposable genetic elements (see Young 1981). It fits that their copy numbers (ranging between about 10 and 100 copies) agree with the copy numbers of repetitive sequences attributed to the transposable part of the genome. The occurrence of such sequences interspread in satellite-like DNA sequences in the Y chromosome (for details see Vogt et al. 1986) indicates the possibility that the Y chromosome may serve as a target for transposable elements. The insertion of a transposable element would in this case not destroy protein-coding regions as easily as in euchromatic chromosome regions. It also would not necessarily affect the postulated protein-binding capacities of Y chromosomal sequences, since even the loss of a binding site would leave sufficient other binding sites to fullfil this function. Insertion of transposable DNA sequences into the Y chromosome would hence not interfere with gene functions to a relevant extent.

In this context it is of interest to recall that multiple copies of a retrovirus genome have been discovered to be deposited in the Y chromosome of the mouse (Phillips et al. 1982). All evidence available implies that the integration of retroviruses bears close relationship to the behavior of transposable elements. Again, heterochromatin may facilitate frequent insertion of movable DNA segments without a high incidence of gene mutations.

It has recently been discovered that the copy number of pepsinogen-coding regions in the human genome varies from individual to individual (Taggart et al. 1985). This gene is located in the centromeric region of chromosome 11. Although a relationship in heterochromatin is not unequivocally established, a position close to or within the centromeric heterochromatin is possible. Similarly, other genes which are subject to frequent changes in their copy numbers are also located close to or within heterochromatic chromosome regions. Well-known examples are the ribosomal DNA, which is usually associated with perinucleolar heterochromatin (see Hennig 1973) or the histone genes in *Drosophila*, which reside in a region conventionally regarded as heterochromatic (Pardue et al. 1973) (cf. also Hilliker and Appels 1980). Although these genes do not belong to the class of transposable genetic elements, they have the common property that they can undergo changes of their copy numbers. The mechanism of this amplification process is still not clearly understood, but it definitely requires a certain structural flexibility of the surrounding chromosome regions.

The potential enrichment of transposable elements in heterochromatic chromosome regions is of interest with respect to processes such as hybrid dysgenesis. It has been demonstrated that hybrid dysgenesis is caused by the activation of transposable elements introduced into a foreign cytoplasm (review: Kidwell 1983). Similarly, reciprocal crosses between related species lead typically to differences in the viability and fertility of both sexes in the offspring generation (see Patterson and Stone 1952). The genomic differences in reciprocal crosses are restricted to the genetic constitution of the offspring regarding their sex chromosomes, if one neglects possible differences in imprinting of the parental genomes and ooplasmic differences. Differences in the composition of heterochromatin of the sex chromosomes may hence play an important role in hybrid dysgenesis.

There is some experimental evidence that incompatibilities in the genomes of hybrid cells induce modifications of heterochromatic parts of the genome. In hybrid cells carrying the genomes of *Nicotiana tabacum* and *N. otophora*, giant blocks of heterochromatin are newly created (Gerstel and Burns 1966).

Our comparative studies on the location of middle repetitive DNA sequences (including transposable fractions such as the 28S IVS sequence) indicated fundamental differences in the distribution of such DNA sequences between the closely related species *D. hydei*, *D. neohydei* and others (Hennig et al. 1983; unpublished data of Hackstein, Hennig and Vogt), although the euchromatic parts of the genomes are cytologically almost indistinguishable (Hennig 1978).

In the context of our discussion it would be highly interesting to know more about the cellular stages where experimental transformation with exogeneous genetic elements of germ cells is achieved (Rubin and Spradling 1982). There is also no information of whether transformation occurs frequently in heterochromatic sections of the genome. Such a location of inserted DNA sequences would probably often not be detected, since position effects would be expected which could prevent the function of such inserted genes.

The evaluation of a function of heterochromatin as target for transposable sequences requires more directed future experimentation. In my opinion, the data available are very indicative of such a function and justify further experimentation to evaluate this possibility.

Concluding Remarks

The proportion of the eukaryotic genome coding for proteins or RNA species with known functions is small. It usually does not exceed 5% of the total DNA in most higher organism. Our knowledge of the biological relevance of the remaining – and overwhelming – part of the eukaryotic genome is almost zero. This is emphasized in the contributions in this book, which consider the present state of understanding the biological role of a distinct part of the genome: the function of major heterochromatic sections. Although it could be argued that cases of elimination of heterochromatin from somatic cells are specialized situations and may concern a particular type of DNA sequences not generally present in many other organism, it is more likely that eliminated chromatin is equivalent to other heterochromatic sections of the genome.

This, however, would not answer the question why the presence of heterochromatin in somatic cells is usually accepted and only in exceptional cases removed by special mechanisms. One can only speculate on the reasons for such a removal from the genome, but there are some features common to genomes which are subject to chromatin elimination. First, the proportion of heterochromatin in germ cells is unusually high. Second, there are large blocks of heterochromatin which reside either in terminal positions of the chromosomes (*Ascaris, Cyclops*), are interspersed in blocks larger than is usual for interkalary heterochromatin (*Cyclops, Stylonychia*), or reside on separate chromosomes (*Sciara*, B chromosomes). Whether these properties enable rejection mechanisms to take place or whether the large amounts of heterochromatin require removal from somatic cells is unclear.

Elimination of heterochromatin can easily be achieved if it is located in separate chromosomes. The presence of large blocks of terminal or intercalary heterochromatin may interfere with an undisturbed function of adjacent euchromatic chromosome regions and therefore require removal with the aid of a specific mechanism. Such mechanism are available in the genome, as is suggested by the observation that highly repetitive DNA fractions are frequently substituted by new clusters of repetitive sequences in evolution. Other indications for amplification and rejection mechanisms are the variable amount of DNA in the genome of flax (Durrant 1981), the substitution of spacer sequences between ribosomal genes (Brown et al. 1972; Brown and Blackler 1972) and, more generally, the distribution of satellite DNA in related species (see for example Walker 1971; Hennig et al. 1970; Brutlag 1980). The general occurrence of multiplication steps in the eukaryotic genome which allow fast amplification of distinct sequences has moreover been demonstrated for several types of DNA sequence. Telomeric sequences expand rapidly (Blackburn and Szostak 1984). Chorion genes can be intrachromosomally amplified in *Drosophila* (Levine and Spradling 1985; Wong et al. 1985). Y-chromosome-specific DNA sequences in *Drosophila* are rapidly substituted in evolution (Vogt et al. 1986). Obviously, special mechanisms must be present which facilitate the removal of blocks of sequences from the genome and their substitution by others.

The presence of large blocks of kinetochore-associated heterochromatin seems, on the other hand, not to be of much disadvantage for the cell. If heterochromatin is restricted to distinct chromosome arms, as, for example, in *Drosophila virilis*, its removal from somatic cells seems not essential. An explanation for the maintenance of such blocks of heterochromatin in somatic cells may be found in the specific structural arrangement compared to adjacent euchromatic chromosome regions.

Unfortunately, the organisms where chromatin elimination can be observed are not very favorable for biochemical studies, except for the ciliates. Their further study will, however, most likely help to reveal potential biological functions assigned to heterochromatic chromosome regions.

Acknowledments. For critical reading of the manuscript and for valuable comments I am very grateful to Drs. Wolfgang Beermann, Mary Ann Handel and Hannie Kremer. The ideas on heterochromatin developed in this article emerged in the course of many stimulating scientific discussions in the context of the research of our group with my colleagues Drs. Rein C. Brand, Johannes Hackstein, Peter Huijser, Hannie Kremer and Peter Vogt.

References

Baldwin L, Macgregor HC (1985) Centromeric satellite DNA in the newt Triturus cristatus karelinii and related species: Its distribution and transcription in lambrush chromosomes. Chromosoma 92:100–107

Beckingham K (1982) Insect rDNA. In: Busch H (ed) The cell nucleus, vol X; 206–269

Beermann S (1959) Chromatin-diminution bei Copepoden. Chromosoma 10:504–514

Beermann S (1977) The diminution of heterochromatic chromosome segments in Cyclops (Crustacea, Copepoda). Chromosoma 10:297–344

Beermann W (1952) Chromomerenkonstanz und spezifische Modifikationen der Chromosomenstruktur in der Entwicklung und Organdifferenzierung von Chironomus tentans. Chromosoma 5:139–198

Beermann W (1984) Riesenchromosomen. Protoplasmatologia, Bd. VI D

Blackburn EH, Szostak JW (1984) The molecular structure of centromeres and telomeres. Ann Rev Biochem 53:163–194

Bostock C (1971) Repetitious DNA. Adv Cell Biol 2:153–223

Boveri Th (1887) Über Differenzierung der Zellkerne während der Furchung des Eies von Ascaris megalocephala. Anat Anz 2:688–693

Britten RJ, Kohne DE (1968) Repeated sequences in DNA. Science 161: 529–540

Brown DD, Blackler AW (1972) Gene amplification proceeds by a chromosome copy mechanism. J Mol Biol 63:75–83

Brown DD, Wensink PC, Jordan E (1972) A comparison of the ribosomal DNA's of Xenopus laevis and Xenopus mulleri: the evolution of tandem genes. J Mol Biol 63:57–73

Brown SW (1966) Heterochromatin. Science 151:417–425

Brown SW, Nur U (1964) Heterochromatic chromosomes in the coccids. Science 145:130–136

Brutlag DL (1980) Molecular arrangement and evolution of heterochromatic DNA. Ann Rev Genet 14:121–144

Cooper KW (1959) Cytogenetic analysis of the major heterochromatic elements (especially Xh and Y) in Drosophila melanogaster, and the theory of "heterochromatin". Chromosoma 10:535–588

D'Amato F (1977) Nuclear cytology in relationship to development. Cambridge Univ Press, London New York Melbourne

Davidson EH (1976) Gene activity in early development, 2nd ed. Academic Press, New York

deLoos F, Dijkhof R, Grond CJ, Hennig W (1984) Lampbrush loop-specific transcripts in Drosophila hydei. EMBO J 3:2845–2849

Diaz MO, Gall JG (1985) Giant readthrough transcription units at the histone loci on lambrush chromosomes of the newt Notophthalmus. Chromosoma 92:243–253

Durrant A (1981) Unstable genotypes. Phil Trans Roy Soc Lond B 292:467–474

Flamm G (1972) Highly repetitive sequences of DNA in chromosomes. Int Rev Cytol 32:1–51

Gall JG, Atherton DD (1974) Satellite DNA sequences in Drosophila virilis. J Mol Biol 85:633–664

Gerstel DU, Burns JA (1966) Chromosomes of unusual length in hybrids between two species of Nicotiana. Chromosomes Today, vol 1. 41–56

Grond CJ, Siegmund I, Hennig W (1983) Visualization of a lampbrush loop-forming fertility gene in Drosophila hydei. Chromosoma 88:50–56

Gutherz S (1907) Kenntnis der Heterosomen. Arch Mikr Anat Entwicklmech 69:491–514

Hamlin JL Milbrandt JD, Heinz NH, Azizkhan JC (1984) DNA sequence amplification in mammalian cells. Int Rev Cytol 90:31–82

Hammond MP, Laird CD (1985) Control of DNA replication and spatial distribution of defined DNA sequences in salivary gland cells of Drosophila melanogaster. Chromosoma 91:279–286

Heitz E (1928) Das Heterochromatin der Moose. Jb wiss Bot 69:762–818

Heitz E (1929) Heterochromatin, Chromozentren, Chromosomen. Ber Dtsch Bot Ges 47:274–284

Hennig I (1978) Vergleichend-cytologische und genetische Untersuchungen am Genom der Fruchtfliegen-Arten Drosophila hydei, D. neohydei und D. eohydei. Ent Germ 4:211–223

Hennig W (1980) Functional units of chromosomes, vol I. Proc XIV Int Congr Genetics Moscow 1:117–128

Hennig W (1973) Molecular hybridization of DNA and RNA in situ. Int Rev Cytol 36:1–44

Hennig W, Hennig I, Stein H (1970) Repeated sequences in DNA of Drosophila and their localization in giant chromosomes. Chromosoma 32:31–63

Hennig W (1972) Highly repetitive DNA sequences in the genome of Drosophila hydei. II. Preferential localization in the X chromosome heterochromatin. J Mol Biol 71:407–417

Hennig W, Huijser P, Vogt P, Jäckle H, Edström J-E (1983) Molecular cloning of microdissected lampbrush loop DNA sequences of Drosophila hydei. The EMBO J 2:1741–1746

Hennig W (1986) Ychromosomal DNA sequences. Trends in Genetics, in preparation

Hennig W (1967) Untersuchungen zur Struktur und Funktion des Lampenbürsten-Y-Chromosoms in der Spermatogenese von Drosophila. Chromosoma 22:294–357

Hennig W, Vogt P, Jacob G, Siegmund I (1982) Nucleolus organizer regions in Drosophila species of the repleta group. Chromosoma 87:279–292

Hennig W (1985) Y chromosome function and spermatogenesis in Drosophila hydei. Adv Genet 23:179–234

Hilliker AJ, Appels R (1980) The genetic analysis of heterochromatin. Cell 21:607–619

Hughes-Schrader S (1948) Cytology of Coccids (Coccoidae, Homoptera). Adv Genet 2:127–203

Huijser P, Hennig W (1986) Ribosomal DNA-related sequences in an Ychromosomal lampbrush loop of Drosophila hydei. Submitted for publication

John B, Miklos GLG (1979) Functional aspects of satellite DNA and heterochromatin. Int Rev Cytol 58:1–114

Jones RN (1977) B chromosomes. Int Rev Cytol 40:1–100

Kidwell MG (1983) Intraspecific hybrid sterility. In: Ashburner M, Carson HL, Thompson JN Jr (1983) The genetics and biology of Drosophila, vol 3c. Acad Press, New York, pp 125–154

Kram R, Botchan M, Hearst JE (1972) Arrangement of the highly reiterated DNA sequences in the centric heterochromatin of Drosophila melanogaster. Evidence for interspersed spacer DNA. J Mol Biol 64:103–117

Kremer JMJ, Hennig W, Dijkhof R (1986) Chromatin organization in the male germ line of Drosophila hydei. Chromosoma, submitted for publication

Levine J, Spradling A (1985) DNA sequence of a 3.8 kilobase region controlling Drosophila chorion gene amplification. Chromosoma 92:136–142

Lifschytz E, Hareven D, Azriel A, Brodsly H (1983) DNA clones and RNA transcripts of four lampbrush loops from the Y chromosome of Drosophila hydei. Cell 32:191–199

Lima-de-Faria A (1959) Differential uptake of tritiated thymidine into hetero- and euchromatin in Melanoplus and Secale. J Biophys Bioch Cytol 6:457–466

Lima-de-Faria A (1983) Molecular evolution and organization of the chromosome. Elsevier, Amsterdam

Lyon MF (1968) Chromosomal and subchromosomal inactivation. Ann Rev Genet 2:31–52

Matthey R (1949) Chromosomes sexuels géants chez un Campagnol, Microtus agrestis L. Experientia 5:72

Matthey R (1950) Les chromosomes sexuels géants de Microtus agrestis L. Cellule 53:162–184

Meyer GF, Hess O, Beermann W (1961) Phasenspezifische Funktionsstrukturen in Spermatocytenkernen von Drosophila melanogaster und ihre Abhängigkeit vom Y-Chromosom. Chromosoma 12:676–716

Müntzing A (1974) Accessory chromosomes. Ann Rev Genet 8:243–266

Nover L, Richter RF, Serfling E (1982) Structure and expression of immunoglobulin genes. Biol Zbl 101:27–55

Ohno S (1973) Evolutional reasons for having so much junk DNA. In: Pfeiffer RA (ed) Modern aspects of cytogenetics: Constitutive heterochromatin in Man. Symp Med Hoechst 6. F.K. Schattauer Verlag, p 69–190 (incl. discussion)

Pardue ML, Birnstiel ML (1973) Cytological localization of repeated gene sequences. In: Pfeiffer RA (ed) Modern aspects of cytogenetics: constitutive heterochromatin in Man Symp Medica Hoechst, 1972. FK Schattauer Verlag, Stuttgart New York

Patterson JT, Stone WS (1952) Evolution in the genus Drosophila. Macmillan, New York

Phillips SJ, Birkenmeier EH, Callahan R, Eicher EM (1982) Male and female mouse DNA can be discriminated using retroviral probes. Nature 297:241–242

Rae PMM (1972) The distribution of repetitive DNA sequences in chromosomes. Adv Cell Mol Biol 2:109–149

Renkawitz R (1978a) Characterization of two moderately repetitive DNA compounds within the β-heterochromatin of Drosophila hydei. Chromosoma 66:225–236

Renkawitz R (1978b) Two highly repetitive DNA satellites of Drosophila hydei localized within the α-heterochromatin of specific chromosomes. Chromosoma 66:237–248

Renkawitz R (1979) Isolation of twelve satellite DNA's from Drosophila hydei. Int J Biolog Macr 1:133–136

Ribbert D (1975) Unterschiedliche Chromosomenmuster von Polytänchromosomen in Keimbahn und Soma der Fliege Calliphora erythrocephala. Nachr Akad Wiss Göttingen II. Math-Phys Klasse 189–192

Rubin GM, Spradling AC (1982) Genetic transformation of Drosophila with transposable element vectors. Science 218:348–353

Rudkin GT (1969) Non-replicating DNA in Drosophila. Genetics (Suppl) 61:227–238

Schimke RT (1982) Gene amplification. Cold Spring Harbor Laboratories, Cold Spring Harbor, New York

Schneider E, Heukamp U, Pera F (1973) Loss of heteropyknosis of the constitutive heterochromatin in specifically activated cells of the thyroid gland of Microtus agrestis. Chromosoma 41:167–173

Schultz J (1947) The nature of heterochromatin. Cold Spring Harb Symp Quant Biol 12:179–191

Scott MP, Pardue ML (1981) Translational control in lysates of Drosophila melanogaster cells. Proc Nat Acad Sci USA 78:3353–3357

Singer MF (1982) Highly repeated sequences in mammalian genomes. Int Rev Cytol 76:67–112

Solter D, McGrath J (1985) Nuclear transfer in mouse embryos – activation of embryonic genome. Abstract Cold Spring Harb Symp Quant Biol 50:8

Sonnenblick BP (1950) The early embryology of Drosophila melanogaster. In: Demerec M (ed) Biology of Drosophila. Hafner Publ Co New York, reprint 1965, pp 62–167

Southern EM (1970) Base sequence and evolution of guinea pig asatellite DNA. Nature 227:794–798

Southern EM (1975) Long range periodicities in mouse satellite DNA. J Mol Biol 94:51–69

Spradling AC, Rubin GM (1981) Drosophila genome organization: Conserved and dynamic aspects. Ann Rev Genet 15:219–264

Stark GR, Wahl GM (1984) Gene amplification. Ann Rev Biochem 53:447–491

Taggart RT, Mohandas TK, Shows TB, Bells GI (1985) Variable numbers of pepsinogen genes are located in the centromeric region of human chromosome 11 and determine the high-frequency electrophoretic polymorphism. Proc Nat Acad Sci USA 82:6240–6244

Tanaka R (1969) Deheterochromatization of the chromosomes in Spiranthes sinensis Japan. J Genet 44:291–296

Tanaka R (1969) Speciation and karyotypes in Spiranthes sinensis. J Sci Hiroshima Univ B Div 2:12, 165–197

Trendelenburg MF, McKinnell RG (1979) Transcriptionally active and inactive regions of nucleolar chromatin in amplified nucleoli of fully grown oocytes of hibernating frogs, Rana pipiens (Amphibia, Anura). Differentiation 15:73–95

Tartof KD, Hobbs C, Jones M (1984) A structural basis for posirion effects. Cell 37:869–878

Varley JM, Macgregor HC, Erba HP (1980) Satellite DNA is transcribed on lampbrush chromosomes. Nature 283:686–688

Vogt P, Hennig W (1986a) Molecular structure of the lampbrush loop nooses of the Ychromosome of Drosophila hydei. I. A Ychromosom-specific repetitive DNA sequence family is dispersed in the loop DNA. Submitted for publication

Vogt P, Hennig W (1986c) Molecular structure of the lampbrush loop nooses of the Ychromosome of Drosophila hydei. II.

Vogt P, Hennig W, Siegmund I (1982) Identification of cloned Y chromosomal DNA sequences from a lampbrush loop of Drosophila hydei. Proc Nat Acad Sci USA 79:5132–5136

Vogt P, Hennig W (1983) Y chromosomal DNA of Drosophila hydei. J Mol Biol 167:37–56

Vogt P, Hennig W, ten Hacken TMM, Verbost P (1986) Evolution of Ychromosomal lampbrush loop DNA sequences of Drosophila. Chromosoma, submitted for publication

Walker PMB (1971) "Repetitive" DNA in higher organisms. Progr Biophys Mol Biol 23:145–190

Wolff BE (1957) Temperaturabhängige Allozyklie des polytänen X-Chromosoms in den Kernen der Somazellen von Phryne cincta. Chromosoma 8:396–435

Wong Y-Ch, Pustell J, Spoerel N, Kafatos FC (1985) Coding and potential regulatory sequences of a cluster of chorion genes in Drosophila melanogaster. Chromosoma 92:124–135

Wu Z, Murphy C, Gall JG (1986) A transcribed satellite DNA from the bullfrog Rana catesbeiana. Chromosoma 93:291–297

Young MW (1981) Repeated sequences in Drosophila, vol 3. Genetic Engineering, pp 109–128

Subject Index